Engineering in Elementary STEM Education

Engineering in Elementary STEM Education

Curriculum Design, Instruction, Learning, and Assessment

Christine M. Cunningham

Foreword by Richard A. Duschl

Published simultaneously by Teachers College Press, 1234 Amsterdam Avenue, New York, NY 10027 and the Museum of Science, Boston, 1 Museum of Science Driveway, Boston, MA 02114.

Cover photo courtesy of Engineering is Elementary, Museum of Science, Boston. Chris San Antonio-Tunis, photographer.

Library of Congress Cataloging-in-Publication Data

Names: Cunningham, Christine M., 1969- author.
Title: Engineering in elementary STEM education : curriculum design, instruction, learning, and assessment / Christine M. Cunningham ; foreword by Richard A. Duschl.
Description: New York, NY : Teachers College Press, [2018] | Includes bibliographical references and index.
Identifiers: LCCN 2017040622 (print) | LCCN 2017054049 (ebook) |
ISBN 9780807776711 | ISBN 9780807758779 (pbk. : alk. paper) |
ISBN 9780807758786 (hardcover : alk. paper)
Subjects: LCSH: Science—Study and teaching (Elementary)—United States. | Engineering—Study and teaching (Elementary)—United States.
Classification: LCC LB1585.3 (ebook) | LCC LB1585.3 .C86 2017 (print) | DDC 372.35/044—dc23
LC record available at https://lccn.loc.gov/2017040622

ISBN 978-0-8077-5877-9 (paper)
ISBN 978-0-8077-5878-6 (hardcover)
ISBN 978-0-8077-7671-1 (ebook)

Printed on acid-free paper
Manufactured in the United States of America

25 24 23 22 21 20 19 18 8 7 6 5 4 3

Contents

Foreword

In 1990, I began a NSF funded project to study 'Portfolio Assessment' strategies in middle school science—Project SEPIA. The first step was to create a curriculum to support complex reasoning and meaning making. We met with teachers and asked them "When were students most interested and enthusiastic?" The teachers quickly agreed that this occurs when the students were involved in building or designing type activities. The context we selected and developed was the building boats. The problem-based Vessels Unit took the 5-day lesson sequence and made it 6-weeks long. We did so to accommodate development of the cognitive and conceptual reasoning needed (1) to build and refine models for the design of a vessel, and (2) to build and refine a causal explanation for explaining in terms of forces—gravity and buoyancy—why the design works.

This was not an easy change for the teachers. One teacher, I recall, at a meeting drew an image of a horse labeled 'Beating a Dead Horse'. It revealed his anxiety about spending 6 weeks on a topic that was once only 5 days. I realized then the challenges with reform of curriculum, learning goals, and assessments were not with students but with teachers. How I wish that I had had access to the tips for teachers and curriculum designers reported here in *Engineering in Elementary STEM Education.*

Fast forward to 2004, chairing the National Research Council consensus study committee that wrote "Taking Science to School: Learning and Teaching Science in Grades K–8" report. As we were engaged in establishing the consensus recommendations and conclusions, I recall one committee member saying "If we are recommending setting the bar this high for students, then we need to think about the implications of where the bar will be for teachers." "Taking Science to School" is the synthesis research report where you'll first find recommendations about science practices, building and refining models, 'Learning Progression' and participatory talk and discourses practices. How I wish, back then, that we had the access to the programs and research conducted by the Museum of Science and reported in *Engineering in Elementary STEM Education.*

Fast forward again to 2012, Next Generation Science Standards, I co-chaired the Earth/Space Sciences Middle/High School writing team. All writing teams were deep into the design of learning goals, standards, and performance tasks that embraced the three dimensions: Practices, Crosscutting Concepts, and Core Ideas. At one point, writing teams were tasked with locating those standards that could foreground engineering, mathematics, and nature of science. Debates about integrating STEM were taking place and feedback was mixed. Some states/districts embraced engineering. Others cautioned adding in engineering along with adopting three dimensional framework changes would be too much for teachers, principals, and educational systems. How I wish, back then, that we had access to the balanced and reasoned perspective about the integration of S–T–E–M presented by Christine Cunningham (VP–Boston Museum of Science) in *Engineering in Elementary STEM Education.*

Are you wondering how to infuse engineering into your teaching and curriculum? Want to understand how you can include the "E" in STEM education? Here's the book for you! Christine Cunningham has written a book that addresses these and other frequently asked questions by teachers and administrators about engineering in the elementary classroom. With the aplomb and persistence of engineers, Christine along with her Museum of Science colleagues have considered the problem, developed a process to solve it, then implemented, evaluated, iterated and adjusted the curriculum, instruction, and assessment models until ultimately finding a solution. *Engineering in Elementary STEM Education* represents a thorough and thoughtful pathway for addressing the integration of E into STEM. And, along the way, you are provided with compelling conversations with teachers with access to classrooms provided via web links.

Christine Cunningham has given us a book that provides rich insights about how to proceed. She entered the engineering education problem space without any preconceived notions except one: that young children are enthusiastic and able learners. Christine has drawn back the curtains of how the Museum of Science's 'Engineering is Elementary' was conceived, designed, developed, and researched. With effective integration of frameworks, Museum advocacy, video examples, and transcripts of lessons, teachers and administrators, and education researchers, too, I might add, get a glimpse into one way of what integrated STEM management looks like and sounds like; all along capitalizing on the R&D efforts she developed and guided.

—Richard A. Duschl,
Waterbury Chair Professor of Secondary Education,
The Pennsylvania State University

Acknowledgments

I never dreamed I'd write a book. To do so, I have drawn inspiration, ideas, and support from educators, colleagues, and family. Over the past 20 years, I have been fortunate to work with thousands of teachers. The talents and dedication of these professionals to create environments in which their pupils learn and flourish truly brighten the future. Their willingness to share their expertise, welcome me and my team into their classrooms, and take risks as they tested our rough drafts of engineering lessons is the bedrock upon which this work has been built. The educators have kept my team grounded in the realities of classrooms as they helped us chart a course that introduced engineering to schools. We have jointly created a vision for engineering with young students. I have the deepest respect and appreciation for our classroom collaborators and all our nation's K–12 teachers—they are a constant source of inspiration and awe. When I need a boost while poring over grant proposals or spreadsheets, I often watch a few minutes of our videos of teachers in action. They help me remember why the daily tasks matter.

One of the best parts of my job is my team. The Engineering is Elementary (EiE) project began at the Museum of Science in Boston with two people and we now number 50. Being surrounded by colleagues who believe in our mission has generated discussion, innovation, and debate that pressure-tested our ideas and morphed them into more rooted and stable forms. I am fortunate that I learn much with and from my team every day! All team members, past and present, helped bring this book to fruition, but some contributed in additional ways: the leadership of Martha Davis, Shannon McManus, and Jonathan Hertel keeps our project focused and running. Their comments on previous drafts of this book have only strengthened it. Cathy Lachapelle, who has worked with EiE almost since its inception, has read the chapters multiple times and provided valuable insights. Chris Gentry, Chris San Antonio-Tunis, Ian Burnette, Corey Niemann, and Lauren Redosh have helped organize the myriad of details that crafting a book entails. The tables, figures, transcripts, images, graphics, and the website are much better because of them. The team's videographers, Kathleen Young, Ellen Daoust, Richard Sutton, and Alex Hennessey, have searched footage from our video library to help bring this book to life. Several other team members have commented on previous drafts and helped clarify my thinking. I am also grateful for the guidance of Annette Sawyer, a fellow museum vice president and honorary EiE team member, whose counsel has helped keep the project humming. I also appreciate the support that Ioannis Miaoulis and the Museum of Science leadership have provided for this work.

Over the past 20 years, it has been gratifying to become part of a network of engineering educators. Colleagues across the nation have pushed my thinking in new ways. My doctoral advisor, Bill Carlsen, took a chance and admitted an unorthodox graduate school applicant. He also first introduced me to engineering education. Conversations with Heidi Carlone, Greg Pearson, Elizabeth Parry, Pamela Lottero-Perdue, Martha Cyr, Stacy Klein Gardner, Chris Rogers, and Cary Sneider have especially helped shape my ideas throughout the years. I have always appreciated their steadfast encouragement.

The EiE project would not exist without the support of a number of generous grants and gifts. They have all supported the idea of engineering education for children and include the National Science Foundation (#01387766, 0454526, 0702853, 1003060, 1220305), the Cargill Foundation, the Gordon Foundation, Raytheon, S. D. Bechtel Jr. Foundation, NASA, Liberty Mutual Foundation, Cognizant, i2Camp, Cisco Foundation, Intel Foundation, National Institute of Standards and Technology, Google Community Grants Fund of Tides Foundation, Massachusetts Board of Education Pipeline Fund, Oracle, U.S. Small Business Administration, SheGives Boston, AIR Worldwide, U.S. Institute of Museum and Library Services, 100Kin10, Hewlett-Packard, MathWorks, Millipore, Dell, Samueli Foundation, and Amazon. The ideas in this book are mine, however, and do not necessarily represent those of these organizations.

This book would not be what it is without the assistance of four editors. Emily Spangler and the team at Teachers College Press escorted this book from idea to reality. Cynthia Berger was an early cheerleader for the book, reading and editing the first drafts. Amielle Major learned of this project on her first day with EiE. She enthusiastically jumped into the project. Her storytelling and writing abilities helped make the book a much friendlier read, I assure you. Finally, my sister Deborah Cunningham contributed her gift with language and her knowledge of K–12 education to the endeavor. I have always admired and benefited from Deb's abilities. Thank you for editing my work for 3 decades—I have learned so much about writing from you!

Without my family, this book and my work would not exist. They have always encouraged me to think independently and take risks. And they have supported me as I have done so. My parents, Jan and Brian, demonstrated for me the joy of learning and the power of education to change lives. They also instilled in me a desire to work for social justice. I am blessed with three amazing siblings, Deborah, Kathryn, and Daniel. They served as my first students as we played hours of "school" many years ago. Since then, they have also taught me much and provided a sounding board for new ideas and challenges. My nieces and nephews—Rebecca, Hannah, Erik, Elizabeth, Benjamin, Nicholas, William, Charlotte, and Margaret—remind me of the unbridled joy, potential, and enthusiasm of young children. They were also some of the very early testers of engineering activities. My stepchildren—Maesso, Colleen, and Patrick—have allowed me to experience the educational system as a parent and deepened my understanding of some of the nuances, inequities, and challenges that still exist. Together, we have explored engineering and problem-solving activities at home and in other educational settings such as museums and camps. Finally, I am continually grateful for the enduring friendship, advice, humor, and ever-present support and love of my husband, Greg Kelly.

Introduction

I've been thinking about ways to improve STEM (science, technology, engineering, mathematics) education for a long time. I started playing "school" with my three younger siblings after I entered kindergarten, and I've been teaching ever since. I didn't have to look far for a role model who cared about STEM subjects: My mother had grown up in a family of builders and machinists who were accustomed to fixing things and designing what they needed. Equipped with their handiness and ingenuity, she added her own interests. In college, she was a math major turned biology major at a time when top-tier colleges were just beginning to open their doors to blue-collar, public school students. Though women were subtly (and not so subtly) encouraged to study subjects like literature, my mother pursued her passions relentlessly, working in science labs throughout her college career. When she met my father, also a biology major, she found a kindred intellect who appreciated her inquisitive mind, her hands-on engagement with the natural world, and her determination to fix every device that broke in our house.

For my part, I have always admired my parents' deep dedication to their children's intellectual development. Our home in a small town in Vermont was full of books and magazines—*Ranger Rick*, *National Geographic World*, series of "how-to" books, biographies of Helen Keller, Harriet Tubman, and Thomas Edison. The entire *World Book Encyclopedia* sat in our living room to be cracked open when we had questions about the world we lived in. My mother, who saw the beauty of science in nature, took us on long hikes in the Green Mountains. With guidebooks in hand, she'd stop and observe an intriguing fungus, a patch of mushrooms, a vocal bird, a dizzying line of scurrying ants, or wildflowers with curious shapes. She always emphasized the power of observation, the knowledge to be gained from watching the hummingbirds gather around the lilac bush outside our kitchen. She modeled a curiosity that is sometimes lost in our educational system. She loved asking questions about why our world operates the way it does: "Why is an icicle shaped as it is?" "How do creatures in tide pools survive?" "What mechanism makes the pendulum swing in the clock?" She was delighted when we asked questions back.

These early, seminal introductions to science and engineering inoculated me against the drudgery of "school science." In the schools I attended, science was barely taught to elementary-age students and taught very traditionally to older ones. Few teachers had degrees or even academic coursework in the science subjects they were teaching, so they relied on "stand and deliver" exercises from a textbook and sometimes just read the book verbatim. Too many of my courses were exercises in meaningless worksheets and grades that had nothing to do with science as I had experienced it. Math was a similar ordeal, where grades mattered more than the substance or its applications in the world.

I dedicated my undergraduate years to developing a deep understanding of the discipline of biology—taking enough science courses to earn both a bachelor's and a master's degree in the subject during those 4 years. Walking home from class one day, a friend told me she was leaving biology to major in philosophy because she "was not allowed to think creatively in science." This jarring sentiment stuck with me, not least because it was expressed by three other people that year. I watched the population of science majors become more and more homogeneous throughout my college career. Students who thought in certain ways persisted; many of the more orthogonal thinkers left. This troubled me—bright, inventive minds were effectively being chased away. Any biologist knows that cross-fertilization and fresh genes (or ideas) ultimately make a population much stronger and able to adapt to new challenges. I suspected that both students and the discipline could benefit from more diverse approaches. It was precisely this creative thinking that advanced science.

I viewed my classes and lab research as tools toward my larger goal, which remained working with public school K–12 STEM education. As a top science student, my decision to pursue graduate school to study *education* was not a popular one—many faculty members at Yale (with the notable exception of my advisor) told me not to "waste my life" studying education. They believed "anyone can teach" and suggested that a more valuable path would be continuing in science. But the possibilities for changing K–12 STEM education had captured my attention and loyalties. Those early experiences with my mother had convinced me that science and engineering were fascinating and that the problem must lie with how they were taught.

For the past 14 years, I've led a team of talented curriculum developers and educational researchers at the Museum of Science in Boston in an effort to create and disseminate engineering curricula for elementary students. Our primary goal has been to identify inclusive design principles and create engineering curricula that support learning for *all* students—including girls, students from racial and ethnic groups underrepresented in STEM, students from low socioeconomic backgrounds, students receiving special education services, and English learners. I chose to focus on engineering over science, technology, or mathematics in part because there were no standardized engineering tests and no expectations for how the subject should be taught. This meant I could think carefully and innovatively about what high-quality activities and instruction might entail.

Hurdles did—and still do—exist. Traditional school instruction, especially with its emphasis on standardized testing, has educated engineering tendencies *out* of students. Too often, students are asked to regurgitate established, "correct" facts or fill in bubbles on exams instead of exercising application of those facts in creative ways that solve problems that are meaningful and relevant to them. Engineering has the potential to engage students as they learn how to tackle problems in a structured yet original ways.

With no expectations for how engineering could be taught to kids, my team and I could investigate the unique opportunities engineering instruction provides. Engineering offers elementary students new ways to understand the human-made (or engineered) world and the natural (scientific) world they inhabit. Engineering activities reinforce a child's critical thinking and problem-solving skills, their creativity, their confidence to innovate, and their ability to communicate and collaborate with their peers. These are all important "21st-century skills" that better prepare students for the complex work and life environments they will join.

Increasingly, teachers and principals across the country are recognizing the value of engineering instruction. Due in part to the advocacy efforts by the Museum of Science, eighteen states and the District of Columbia have adopted the Next Generation Science Standards (NGSS Lead States, 2013), which include engineering, and many other states include engineering in similar or significant ways in their state science standards. However, many eager educators still need support to incorporate engineering in their curricula. Teachers tend to know little about engineering and feel unprepared to lead engineering lessons. Administrators are unsure how to integrate engineering and support teachers' professional development. Interested parents, community leaders, and teacher educators need to know more about the benefits of elementary engineering before they can advocate for engineering instruction in their children's classrooms.

If you have ever wondered how to infuse engineering into your teaching, school, or district curriculum in age-appropriate, inclusive, and engaging ways, this book is for you. It aims to help anyone involved in elementary education—classroom teachers, STEM specialists, and principals working in schools; preservice teachers and education faculty at colleges and universities; scientists, engineers, and organizations involved in STEM outreach; professional development providers working with teachers; school board members and policymakers; parents and community leaders—understand how you can include or support the inclusion of the E in STEM education in ways that align with the realities of classroom life.

This book explains why engineering should be included as a part of an elementary (and K–12) education. It offers organizing structures and frameworks to help you understand what engineering is and how to integrate it into your classroom. I draw from the work of an interdisciplinary curriculum design team that has spent thousands of hours reading literature on how students learn, observing engineering activities in elementary classrooms, and talking with teachers and students to understand how engineering could be taught to elementary students. We followed their advice and created a set of print, photographic, and video curricular materials, educational vignettes, case studies, and assessments that would be most valuable in supporting them as they learn to implement engineering in elementary classrooms. These concrete, real-world examples help educators visualize high-quality elementary engineering instruction and demonstrate how theoretical design principles apply to real students in real classrooms. Throughout this book, I'll describe scenes from classrooms, and provide links to the videos and other resources, so that as you read, you can use a computer, tablet, or smartphone to see for yourself what elementary classroom engineering looks like. I'll signal such resources by providing a URL for the resource. The resources are all available at eie.org/book. Individual resources are coded by chapter and resource number. For example, eie.org/book/3b denotes the second resource in the third chapter. After reading drafts of this book, classroom teachers also requested that I create a set of Focus Questions they could use to guide individual reading of the book or support teacher study groups. The questions can also be downloaded on the resource page (eie.org/book). I hope you find these external resources helpful.

I designed this book to address the most frequently asked questions about engineering in the elementary classroom. It has three sections. The chapters in the first section provide an overview and an introduction. I begin with reasons for including engineering at the elementary level. I unpack "STEM," describe students' baseline conceptions of technology and engineering, and offer activities that can help students construct more accurate understandings of these ideas. Rich vignettes from actual classrooms illustrate what engineering looks like with young students.

The second section of the book starts by examining the Next Generation Science Standards with an engineering lens. Then I describe some engineering habits of mind and design elements that make K–12 engineering curriculum inclusive and effective for all students. The section ends with an exploration of some common concerns teachers have voiced about implementing engineering in their classrooms. If you're a curriculum developer, classroom teacher, school or district administrator, or outreach specialist, the information, frameworks, and examples in these chapters can guide your engineering education efforts and decisions.

The last section provides an overview of my team's research results. What impacts can engineering have on students, teachers, and schools? Our research suggests that our approach to elementary engineering can foster deeper understandings of engineering and technology as it strengthens students' science knowledge. Teachers and students alike develop familiarity with and build affinity for problem solving and engineering. I hope the research findings will help as you advocate for including engineering and consider how to structure and support engineering experiences. Elementary engineering education is still in its infancy—although we have learned much, there is still so much more to understand. The concluding chapter considers how we can continue to improve engineering education.

If approached thoughtfully, engineering can promote cross-curricular connections and project-based learning without adding too significant a burden to your workload. Engineering can serve as an integrating subject by asking students to apply science and mathematics content and skills as they solve a real-world problem. In addition, engineering can also interface with literacy and social studies, by having students read and write for a purpose. I hope that by reading this book, you will be able to create educational experiences that help your students better understand and engage with the human-made world in which they live.

Part I

INTRODUCING ENGINEERING TO ELEMENTARY EDUCATION

CHAPTER 1

Why Make Engineering Part of Elementary Instruction?

During several stimulating years of graduate school at Cornell, I read academic works describing and critiquing the ways that science really works. I knew school science usually did not accurately capture how science was practiced. And it certainly did not often resemble the engaging and fun projects I had experienced outside the classroom! Focused on making school science better, I worked with a group of science education faculty, science faculty, science outreach specialists, and teachers who were committed to developing resources that supported educators in offering more authentic science experiences. We created environmental science curricula and professional development for teams of middle and high school science, math, and technology teachers.

Each year, as part of our program, we developed one novel engineering challenge and held an Engineering Design Congress on campus where student teams from each school shared their engineering designs. One year, a 5th-year senior student participated. He lived on a small rural farm where he often helped his family fix machinery and tools. He had failed English class the previous year, and his teacher, who wanted earnestly for him to succeed, was still trying to figure out ways for him to do so. That year, the challenge we developed was to engineer a device to collect water samples. The specifications detailed that the device needed to be able to collect a sample of water 4 feet from the shore and 6 inches under the top of the water. The challenge hooked the student. He assumed a leadership role on his team. He regularly used the classroom computer to search for, and read, information online to inform his group's design. The opportunity to build and test a collection system played to the student's strengths. This task was not as daunting to him as most school assignments.

Students presented their designs to their peers, who rated and reviewed them. As I watched, I noticed whose designs were ranked highest by peer review and I spoke with the participating teachers. A theme began to emerge—students who were not traditionally "good at school" or who had checked out were re-engaging during the engineering lesson. That year, that same student's group had the top-ranked design. The teachers attending the congress remarked upon the power of the experience to motivate student learning and performance overall and, in particular, shake up classroom hierarchies about who was good at science as well as reshape students' perceptions of their abilities.

The design challenges got me thinking more deeply about engineering as a discipline. How was it different from science (or other subjects)? I delved into literature about underrepresented groups in science and engineering. Some of the attributes of science set forth to explain why women and other underrepresented groups left the discipline were not present in engineering. For example, professional scientists often study small, isolated elements of systems; seek a single "correct" explanation with little room for self-expression; and work fairly independently of the larger social and political context. They might, for instance, study the interaction of a cluster of neurons in frogs' brains. Engineering often allows for more varied solutions, better allowing for different opinions and perspectives, and engineering usually needs to consider the larger context in which the technology will be used. I wondered if these features could be leveraged to invite participation in engineering and science by a wider range of students, particularly those who are underserved, underrepresented, and underperforming—girls, students from races and ethnicities underrepresented in STEM, students from low socioeconomic backgrounds, students receiving special education services, and English learners.

I turned my attention to how engineering might be integrated into science, mathematics, and technology classrooms in ways that interested and engaged *all* students, especially those traditionally underrepresented in these disciplines. For the first couple of years, I worked at Tufts University Center for Engineering Education and Outreach. I focused on the grades I had worked with previously and developed engineering curricula and professional development for middle and high school teachers. However, our work pointed to the fact that engineering education needed to start even earlier—with elementary students.

Although precollege engineering had started to enter some high school and middle school classes by early 2000, engineering with young children was virtually unheard of. The idea was so innovative that many people thought it was crazy, in part because they did not understand what it might look like. Some engineers vehemently argued that their discipline could not be possibly be taught to grade-school students. Similarly, educators could not fathom how 7-year-olds could engineer. Shortly after I joined the Museum of Science in Boston, which had launched efforts to implement engineering at the K–12 level. I started to actively develop an engineering curriculum for elementary students, and I mentioned in a conference presentation that I was embarking upon engineering at this level. An educator raised his hand during the question period and exclaimed incredulously, "What?! You're going to do calculus with 2nd-graders?"

This wasn't the only incident. Over and over, I got the message: "Teach engineering in elementary school? Never gonna happen." At the time, many educators and engineers had a limited vision of what engineering is and a difficult time envisioning the early development of engineering skills. This motivated me to think about what age-appropriate engineering might look like and demonstrate that students can do it. I believed there could be a developmental progression for engineering, just as there are progressions for science, mathematics, and reading.

STUDENTS' CONCEPTIONS OF TECHNOLOGY AND ENGINEERING

Before we could create effective instructional materials and activities, however, we needed to understand how children were thinking. What did children think engineering and technology were? In 2002, as my team and I started to think about introducing engineering to students, we did what good developers do and scoured the educational literature for papers that would enlighten us about children's conceptions of engineers, engineering, and technology. We needed to understand what children know, and what misconceptions they have. Our search yielded little previous work in this area, so we decided to conduct our own research.

How do you measure a student's understanding of technology and engineering? We started by designing a survey. It asked more than 700 students in grades K–12 to draw a picture of an engineer at work and describe their picture in words. Figure 1.1 provides some representative depictions. We coded students' responses and learned that many students hold a number of stereotypes and misconceptions—for example, engineers are construction workers who build structures, bridges, and roads; computer technicians; train conductors; or auto mechanics. From a child's perspective, these assumptions make sense. For example, the word *engineering* contains the word *engine*. Perhaps the most common things that children encounter with engines are cars. Cars have engines, so engineers must fix cars.

After analyzing the data from our initial study, we observed some troubling commonalities. First, students don't have a very accurate or complete understanding of what engineers do. Students did not have an awareness of the range of fields that engineering encompasses; the fields of engineering they depicted were limited. If students' perceptions of engineering are inaccurate or incomplete, this could affect their understanding of the role of engineering in their lives and their interest in engineering as a career. Second, the types of work that students associated with engineering were commonly male-dominated occupations. At present, most construction workers, auto mechanics, and computer technicians are men. So, these (incorrect) perceptions could

Figure 1.1. Students' Drawing of Engineers

influence girls' affinity toward the discipline. Third, we suspected from this first study that in thinking about engineering, students considered the object and not the action. That is, any kind of work conducted on a building or a computer was "engineering," regardless of whether it entailed designing, constructing, or painting the building.

But we also knew that the design of our first research instrument was problematic. We had only asked students to generate one drawing. Perhaps they really knew more than they were able to draw. Perhaps they were choosing more "stereotypical" responses, but their knowledge included many more (unrepresented) fields of engineering work. To more deeply and systematically probe what children knew, we developed a better research instrument, a second questionnaire called "*What Is Engineering?*" This instrument, which we have revised five times since we first used it, provides images and/or descriptions of tasks and asks students to indicate whether or not an engineer would do the task at hand. Students also respond to the open-ended question "What is an engineer?" (View this instrument at eie.org/book/1a.) We've had more than 30,000 elementary-age students complete this instrument. Our findings (see Figure 1.2) reveal similar patterns among students from both sexes, all geographic regions, and racial/ethnic groups: Students tend to focus on the descriptor *noun* and not the *verb*. Instead of selecting items where people are designing or developing new technologies, students select those responses that focus on any type of work done with electronics, cars, or buildings. For example, students favor such items as "install wiring" or "repair cars" instead of items like "design ways to clean water" or "develop better bubble gum."

Figure 1.2. Students' Responses to *What Is Engineering*? (Baseline)

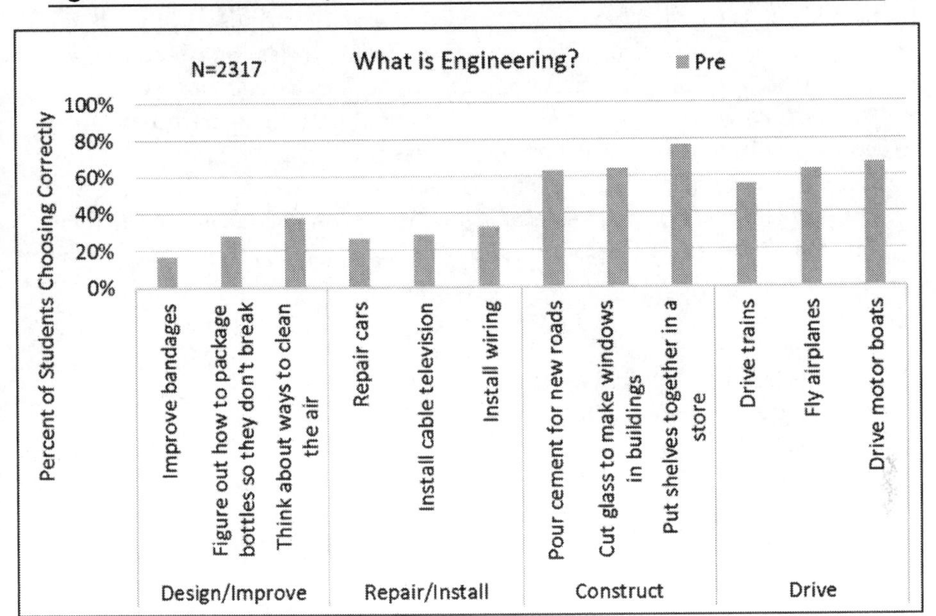

Students focus on the subject of the work, not the type of work being done (Lachapelle, Phadnis, Hertel, & Cunningham, 2012).

Students' open-ended responses echo these results—for example, they indicated:

- An engineer is a person that build [sic] constructs buildings, cars, electricity, and stuff that involve building.
- An engineer is a person who works on different things like engines and electronics.
- An engineer is a person who works on cars. Takes cars apart. They work on stuff with moters [sic]. That is what an engineer is.

Our research on children's ideas of engineers, our conversations with students and teachers, and research by others suggests that children also have misconceptions about what technologies are. Ask a child to name a technology. Most likely, he or she will name an item that uses electricity—often a smartphone, tablet, computer, or some other electronic device. This is not surprising—it's how the word *technology* is most commonly used in the media and in everyday conversations. However, we wanted to understand more deeply what kids knew—and did not—about technology.

So, my research team developed the *What Is Technology?* instrument. It asks students to indicate which items are technology and how they know something is a technology. (View this instrument at eie.org/book/1b.) The results from the 30,000 elementary students we've surveyed indicate that students identify technology with items that run on electricity and power (Lachapelle, Hertel, Jocz, & Cunningham, 2013). Students also correctly identify that things from nature, such as birds and trees, are not technologies. However, they are unlikely to view simpler, commonplace, nonelectrical and nonmechanical items, such as a broom, cup, or shoe, as a technology (see Figure 1.3). Slightly more students identify mechanical but nonelectrical objects, such as a bicycle, piano, or windmill, correctly as a technology.

Students' responses to why something is a technology demonstrate their connection between technology and power and electricity. For example, students wrote:

- I know that an object is technology because they have to have elerecty [electricity] or energy in it.
- A technology is light-tricity.
- For me Technology is like a phone, TV, radios, cellphones, and many things that work with powers.
- And if it gets wet, sparks come out.
- If it uses complicating microchips, engine, computering, and more things that mostly use electricity.
- I know if something is technology I see if the object runs on electricity and I don't have to do any work.

Our research supports the need for elementary engineering education to help students develop more accurate, complete understandings of what technology is. They need to understand that it includes the nonelectrical and nonmechanical items that pervade our lives, and encompasses not only objects, but also systems and processes. Helping students develop more comprehensive ideas about engineering and technology should be one of the foundational principles that STEM elementary instruction nurtures. Students need to understand the wide range of problems that engineers address—not only civil, mechanical, and computer engineering problems, but also biomedical, environmental, chemical, green engineering challenges, and so forth. Furthermore, we need to help young students understand that engineers *design* new technologies of all sorts, not just build or fix them. Children, even young children, *can* engineer. Children *should* engineer. Here are eight reasons why.

WHY ELEMENTARY ENGINEERING?

Engineering Helps Children Understand and Improve Their World. We spend over 98% of our time interacting with products of engineering. To understand the world they live in, children should recognize that engineers helped create almost all of the objects that surround them, and they should understand how

Figure 1.3. Students' Responses to *What Is Technology?* (Baseline)

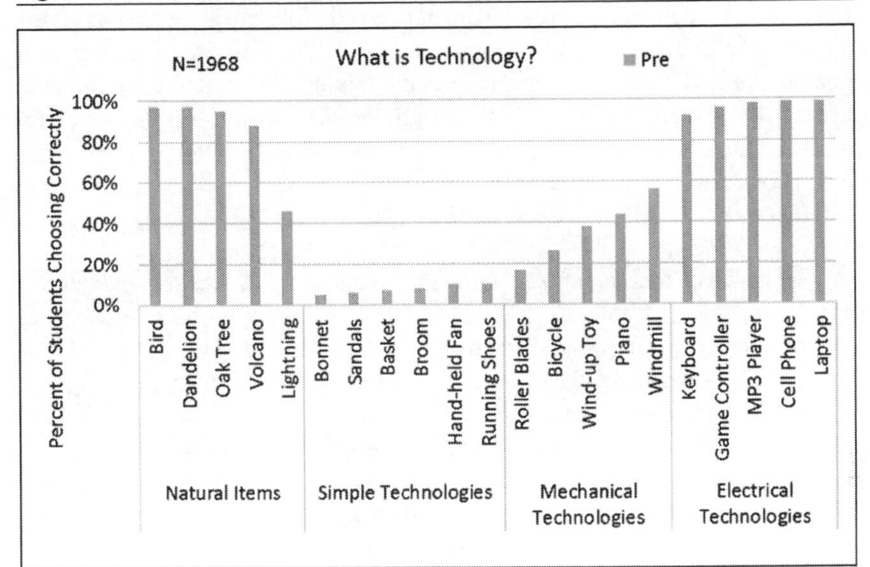

these technologies come into existence. Increasingly, our society depends on engineering and technology. All citizens need a basic understanding of these disciplines so they can make informed decisions as they weigh the benefits, consequences, and costs of new technologies.

Children take for granted that unspoiled food appears in boxes on supermarket shelves, lights turn on when you flip a switch, and waste disappears when you flush the toilet. Understanding that engineering enables these technologies and phenomena allows kids to recognize that we can design and shape the world around us, advocate for new or different solutions, and support initiatives or ballot measures that fund or disseminate availability of life-altering products.

Engineering Fosters Problem-Solving Skills and Dispositions. Human beings are inclined to solve problems and modify their environment. Children informally engineer as they explore the world around them. Classroom activity can harness these tendencies, scaffold them, and help students develop mindsets and strategies to tackle increasingly complex problems. For example, problem solving often requires breaking a problem into smaller parts, persistence, and the ability to learn from failure. These attitudes can be nurtured as students engage in structured engineering design. Throughout their lives, children will encounter various kinds of problems. Cultivating strategies that build children's confidence and perseverance and that help them generate and evaluate potential solutions will serve them well in school and in life, regardless of the

career path they choose. We cannot imagine many of the problems or opportunities that the next generation will engage with. There will be new communication methods, forms of transportation, and technologies to diagnose and treat diseases. The amount of available content and information is exploding at a rate that makes mastery of all the relevant facts impossible. Instead, to equip our students for the future, we should ensure that they develop problem-solving processes and skills such as creativity, collaboration, and communication (Partnership for 21st Century Skills [P21], 2009).

Engineering Can Increase Motivation, Engagement, Responsibility, and Agency for Learning. Students are motivated and engaged by projects that are relevant, meaningful, and open-ended (Blumenfeld et al., 1991; Miaoulis, 2010). Well-designed engineering activities can be all three. After all, students are affected by engineering daily, engineering will shape the world they live in, and there are many possible solutions for any given engineering challenge. The opportunity to generate original solutions to real-world problems can pique students' interest and motivation. Kids will devote hours to creating a water filter that will clean dirty water, an earthquake-resistant building, or a package for humanitarian airdrops for an area isolated by a natural disaster—designing and redesigning it to make it better meet the criteria. Students connect to the engineering contexts or to the clients who will benefit from the solutions, can see the potential impact of their efforts, and come to care about the impacts of their technologies. Engineering tasks also allow students to assume more responsibility for their learning. Students develop ownership of the original designs they create. Because their design is unique, they need to carefully consider next steps for their projects, what they need to know, and claims they can make—they cannot turn to other groups or the teacher for information.

Engineering Can Improve Math and Science Achievement. As students use and manipulate knowledge, they understand it in deeper ways. It is not surprising that when students engage in engineering projects that ask them to apply science knowledge, their understanding of science improves (Fortus, Dershimer, Krajcik, Marx, & Mamlok-Naaman, 2004; Mehalik, Doppelt, & Schunn, 2008). Designing really effective but environmentally friendly insulation for a solar oven rests on an understanding of the properties of insulators. Engineering an alarm circuit that can activate both a light and a buzzer requires an understanding of the differences between parallel and series electric circuits, and creating an effective package for a plant means that students understand the basic needs and functions of plants and their structures. Science concepts that were only "academic" in nature can suddenly become useful when they can inform or improve the performance of a design.

Similarly, engineering projects also invite teachers and students to use mathematics in meaningful ways. Students measure, use mathematical functions, graph, and interpret and analyze their data as they design, construct, and improve solutions. How do you ensure that the parts of a system all fit together accurately? How do you "know" that one design performs better than another? How can you figure out trends in the data you collect? Mathematics-in-action can assist with all these challenges. Relying on math and science to help solve problems in a context can help students understand the connections between these disciplines, as it provides a motivation for understanding and accuracy in these domains.

Engineering Can Increase Access to STEM Careers. Learning about engineering can help students develop accurate understandings of the types of work engineers and technicians do and the role they play in shaping the world around us. Engaging in engineering can build students' confidence in their engineering abilities, develop their agency and identity as engineers, and expose them to engineering disciplines that could be possible career choices (Kelly, Cunningham, & Ricketts, 2017). By referring to the work they do as "engineering" and the students as "engineers," teachers can introduce their students to these career options and support students' proclamations that they "are engineers."

Research suggests that many scientists and engineers began to develop interest in these careers during elementary school (Maltese & Tai, 2010). By middle school, girls and minorities often start to show declining interest in science and math (Catsambis, 1995). Introducing all students to engineering early in their schooling allows them to consider it as a possible career choice before they develop stereotypes about STEM or who is "good at" these subjects. When they actively engage and affiliate with engineering, they can consider this as a possible future. Students who are excited about engineering in elementary school will still need high-quality experiences in middle school and high school, and will need to enroll in appropriate science and mathematics courses to prepare them. Exposure to engineering by *all* students can make them aware of future possibilities and inspire them to consider these as a career.

Engineering Promotes Educational Equity. Engineering is a new discipline at the K–12 level. Thus, engineering can be implemented from the start using inclusive practices that engage diverse students. The ground rules for participation are different. Innovation, originality, and out-of-the-box thinking is prized. Tasks are tied to a client and a context that makes it real. You can show what you think, not only read or write about it. You are supposed to fail and then try again. Teamwork and collaboration are critical. There is no single correct answer. Redefining the academic playing field can also level it, especially for students who did not thrive in more traditional settings. Carefully designed engineering experiences have the potential to change classroom dynamics, students' perceptions of themselves or their peers and their abilities, and teachers' perceptions of their students. Dr. Heidi Carlone at the University of North Carolina at Greensboro studies issues of equity, access, and identity. Her work with elementary engineering suggests that traditional classroom hierarchies of who is "smart" are often disrupted during engineering—a much more diverse range of students self-identify or identify their peers as "smart engineers" (Carlone, Lancaster, Mangrum, & Hegedus, 2016). (View a short video about this research at eie.org/book/1c.) Students who are not the highest-performing students academically often are recognized as "good engineers." Perhaps this is because engineering is a "new" discipline that carries little baggage; and perhaps because engineering requires different traits, it has the power to engage *all* students in meaningful and transformative ways.

Engineering Has the Potential to Transform Instruction. Working with teachers and students across the country, I have seen how introducing engineering can prompt a teacher or a school to rethink instruction more generally. High-quality engineering challenges involve hands-on, project-based, open-ended experiences with no single correct answer. Removing the pressure to guide all students to the correct response encourages teachers to consider how to restructure their role so they can support student groups as they pursue an original solution. Many teachers report that the experience encourages them to "open up" other subjects they teach and try more open-ended challenges. Teachers also value the cross-disciplinary nature of engineering projects—students can draw from and connect science, mathematics, English language arts, social studies, and communication. As Chapter 7 describes, the powerful student responses, their engagement, and the quality of their work have encouraged teachers and schools to rethink how they structure lessons and even the school day.

Engineering Is Included in State and National Standards. Because of the reasons listed above, many state and national standards now incorporate engineering. The Next Generation Science Standards include engineering as well. As states adopt these standards, they should be introducing engineering to their classrooms. Other states have also modified their science and technology standards to incorporate engineering. Meeting such standards requires that students engage with engineering ideas and activities.

Exposing students to engineering can change their understanding of the world around them; their perceptions of their abilities; their engagement with school, projects, or academics; and their future possible career choices.

Engineering with young students allows us to harness the kinds of problem solving and construction that they naturally engage in as they play and construct. As students grow and gain more cognitive, physical, and social maturity, we can scaffold their engineering activity to introduce more "engineering" tools, such as math, science, justification for design decisions, and diagrams and models.

Teaching engineering to students at an early age does make sense. We live in a designed world. Almost everything we see has been created by humans to meet a need or desire: shoes, sticky notes, coffee mugs, toothbrushes, bandages, bicycles, office buildings, and smartphones. These inventions were developed by engineers to solve problems. Throughout history, humans have identified problems and opportunities and modified the world around them.

These problem-solving proclivities start early, much earlier than many of us might expect. Children naturally engineer. Watch young children at play and you'll see them engage in engineering behaviors. They build bridges, houses for dolls, forts, or sandcastles, watch them topple, and then rebuild them. They take devices apart to figure out how they work—sometimes without parental permission. They imagine fantastical solutions for new devices or technologies. Such inclinations offer a great base upon which to build children's understandings of engineering and the engineered world around them. As children interact with the world, it's clear they are building a foundation for future understandings: how materials behave; how technologies are designed, organized, resourced, and constructed; and how engineering affects people, society, and the environment. They are also building a vision of themselves as engineers—an identity that may prove important for their course and career choices down the road.

CHAPTER 2

Putting the Spotlight on the "E" in STEM Education

Although science, math, and engineering disciplines have separate goals and are often taught quite independently in schools, in the real world, these disciplines are closely connected. Many scientists and engineers move back and forth among all these pursuits as they work. Often, scientists use instruments to measure or study physical phenomena, and sometimes they may need to design and construct new instruments tailored to their research—so they engineer the cutting-edge tools they will use to collect their own scientific data. For example, geophysicists researching seismic activity at the bottom of the ocean to better predict earthquakes and tsunamis may need to engineer the detectors they will then use to collect wave data. This new technology may generate more accurate data and advance the related scientific theories.

Similarly, developing effective solutions often requires engineers to use scientific information. For example, biomedical engineers developing a new type of bandage to care for burn wounds need to use their scientific knowledge to do so. They research and draw on their knowledge of biology and chemistry to better determine the most effective ways that bandages can aid healing. As they conduct investigations, engineers develop more advanced solutions for their engineering problems. At the same time, as engineers work within and examine biological systems, they may also advance scientific knowledge of this natural system, thus contributing to the description of the natural world in the process. These are only two examples of how science and engineering overlap. There are even more ways that science, engineering, and math overlap. For example, as engineers and scientists create or use scientific and engineering models, they use mathematical principles and ideas to describe their systems and achieve their goals.

Despite how interconnected these disciplines are in the real world, in our schools these disciplines are usually taught separately. The term *STEM education* is riding a wave of popularity right now. New STEM magnet and charter schools are popping up and existing schools are being renamed to include STEM in their school names. Schools tout their STEM programs and projects in newsletters and websites. STEM-themed after-school clubs and summer camps are also popular. Industry and corporate leaders broadcast the need for a STEM-literate workforce. And politicians call for the need to change the educational system. There is currently much attention and emphasis on STEM, to such an extent that it is becoming an overused and vague term. It's becoming less clear what STEM actually is and how it might be included in the elementary classroom. Regardless of how STEM subjects are taught, educators should help students understand how all the STEM disciplines are interconnected.

This chapter unpacks the commonly used *STEM* acronym. It begins by briefly considering the individual components and how they interact. Because engineering education aims to help students generate accurate and robust understandings of engineering and technology, I include descriptions of lessons that can actively help students construct such understandings. The last section of the chapter explores the idea of integrated STEM and offers an example of how such instruction might be structured.

STEM DISCIPLINES AND DEFINITIONS

The acronym *STEM* is relatively recent. Before considering the four disciplines it includes as a whole, let's think about each individually. What are they? What do they aim to do? How are they similar, different, and overlapping? There are many definitions and examples for each of these disciplines. In this section, I offer brief definitions that are appropriate for elementary-level educators and their students. These functional definitions rest heavily on the goals of the disciplines. As teachers and their students engage in STEM activities and move back and forth between the disciplines, opportunities should arise that help students develop understandings of the relationships and the differences between the disciplines.

> **Science** is a body of knowledge about the physical and natural worlds. Scientists seek to describe, explain, and predict the natural world and its physical properties.

For example, scientists have predicted and explained when and why we see phases of the moon, determined which genetic changes produce a disease, and created models and equations that describe how fast an object falls.

> **Technology** is the body of knowledge, artifacts, processes, and systems that results from engineering. Technologies are produced by humans to solve problems or meet needs and are the products of the process of engineering.

The telephone, coordinated traffic lights and patterns, and the leg brace developed by engineers are all technologies.

> **Engineering** is the application of knowledge to creatively design, build, and maintain technologies. Engineers seek to optimize solutions for problems, needs, and desires while considering resources and various constraints.

One of the most memorable definitions of engineering is also one of the shortest. William Wulf (1998), a previous president of the National Academy of Engineering, offered this terse definition: "design under constraint."

For example, engineers have designed tools that allow people to communicate over long distances, systems that maximize the flow of vehicles through a city, and devices that allow people without the full use of their limbs to stand.

One thing to note is that technology is an "outlier" in this STEM group. Over time, the term *technology* has been used in different ways by different groups. Some use the word to describe information technology or digital technology. Some schools, particularly middle schools, have "technology" classes that evolved out of the industrial arts or vocational technology programs of yore. I consider technology the product of engineering, not a stand-alone discipline. In this book, when I use the term *engineering*, it encompasses both the process of engineering and the resulting products (technologies).

> **Mathematics** is the "science of numbers, quantities, and shapes and the relations between them" (Mathematics, n.d.). Mathematics uses numbers and symbols to describe relationships between concepts. Many other disciplines, including science and engineering, often use "the language of math."

For example, mathematicians have discovered and defined mathematical constants such as π (pi) and they have developed the formulas that describe the circumference and area of a circle. They have created imaginary numbers and articulated relationships between the sides and angles in various shapes. Table 2.1 captures some of the critical features of the STEM disciplines.

My point in offering these definitions is not to suggest that students should memorize or be accountable for them. Rather, the definitions provide a way for explaining concisely what each of the disciplines aims to do. As students engage in STEM activities, pausing to point out when they are doing or using science, engineering, or mathematics can help them understand the utility and interconnectedness of these related disciplines.

Table 2.1. Critical Features of Science, Engineering, Technology, and Mathematics

Science	Technology	Engineering	Mathematics
. . . is the body of knowledge of the physical and natural worlds.	. . . is the body of knowledge, systems, processes, and artifacts that results from engineering.	. . . is the application of knowledge in order to design, build, and maintain technologies.	. . . is the science of numbers, quantities, and shapes and the relations between them.
. . . seeks to describe and understand the natural world and its physical properties.	. . . can be used to describe almost anything made by humans to solve a problem or meet a need.	. . . seeks solutions for societal problems, needs, and wants.	. . . is a language of science and engineering.
. . . uses varied approaches—scientific methods such as controlled experiments or longitudinal observational studies—to generate knowledge.	. . . results from the process of engineering.	. . . uses varied approaches—for example, engineering design processes or engineering analyses—to produce and evaluate solutions and technologies.	. . . uses numbers and symbols to describe the relationship between concepts.
Scientific knowledge can be used to make predictions.	Technologies are anything made by humans to fill a need or desire.	Engineering aims to produce the best solutions given resources and constraints.	Mathematics creates expressions that describe relationships between concepts and the patterns and structures found in nature.

STEM IN K–12 SCHOOLS—TOGETHER BUT NOT (YET) EQUAL

The compilation of the four disciplines—S, T, E, and M—into a new acronym, *STEM*, does not mean that they are equal partners in elementary classrooms. These disciplines have long held different statuses within elementary school settings. Traditionally, K–12 schools have had a strong focus on mathematics with a sprinkling of science. Today, most schools and teachers still attend only to mathematics, and to a much lesser degree, science. But this is starting to change.

Mathematics, once one of the 3Rs ('rithmetic), remains a heavily tested subject at the elementary level. The 2001 Elementary and Secondary Education Act (ESEA), also known as the No Child Left Behind law, and the 2015 Every Student Succeeds Act (ESSA) require schools to test students annually in mathematics in grades 3–8 and once in grades 9–12. School scores on state math tests are examined, and school autonomy and funding are linked to the outcomes. Not surprisingly, almost all (99% K–3, 98% 4–6) teachers teach mathematics all or most days of a week, spending on average 54 minutes per day in grades K–3 and 61 minutes in grades 4–6 (Banilower et al., 2013).

Whether and how much science is taught in elementary and secondary classrooms varies greatly. If you're a teacher or administrator, you know that although all states have science standards, the testing of science topics is less frequent. The ESEA and ESSA both require testing science only once during each of three grade spans: 3–5, 6–8, and 9–12. More important, students' scores on science tests are not considered in decisions about school autonomy and funding. With this in mind, it's not surprising that elementary schools dedicate fewer hours to science than to mathematics instruction. According to a recent study, "Many elementary classes receive science instruction only a few days a week or during some weeks of the year" (Banilower et al., 2013, p. 53). Only 20% of grade K–3 classes and 35% of grade 4–6 classes receive science instruction most or all days. On average, teachers spend 19 minutes on science instruction each day in K–3 classes and 24 minutes in grades 4–6 (Banilower et al., 2013). Over the past 20 years, the amount and percentage of time devoted to science instruction has declined, while instructional time has risen for other core subjects (Blank, 2012).

Engineering and technology are pretty new to the elementary level; most schools still ignore these subjects even today. In the early 90s teaching engineering and technology in elementary schools was inconceivable to some. In 1994, Ioannis Miaoulis, then Dean of Tufts University's School of Engineering, and Professor Chris Rogers founded the Center for Engineering Education and Outreach (CEEO) which aimed to introduce engineering into the lives of young children, starting in kindergarten. During his tenure at Tufts, Miaoulis began and spearheaded engineering standards reform efforts that would result in Massachusetts becoming the first state to include engineering in its K-12 state science and technology standards. In 2001, Massachusetts became the first state to test engineering as part of its state science exams for 5th, 8th, and 10th grade (Miaoulis, 2010).

Miaoulis assumed the presidency of the Museum in 2003 to reposition it as the national leader in engineering education for children in both formal and informal education, and he continued his engineering standards reform efforts. When he arrived, he built the National Center for Technological Literacy (NCTL), with the mandate to introduce engineering in K-12 schools and informal education though advocacy, curriculum development, and professional development of teachers. Because of the NCTL's continued advocacy efforts, engineering has been incorporated into individual state standards, national standards, legislation, and assessments, such as the Framework for K-12 Science Education, Next Generation Science Standards, and National Assessment for Educational Progress (Miaoulis, 2017). These milestones have jumpstarted conversations about teaching engineering at the elementary and secondary level. These milestones have jumpstarted conversations about teaching engineering at the elementary level. Even as more and more educators learn more about the benefits of elementary engineering education, teachers still need a lot of support to become knowledgeable about and comfortable with engineering. Currently, only 9% of elementary teachers indicate that they feel "fairly well" or "very well" prepared to teach engineering (Banilower et al., 2013, p. 24). This is not surprising—engineering is not something that they would have studied during their own K–12 schooling or as part of their teacher education preparation programs. As STEM percolates into schools and classrooms, teachers need to think carefully about how engineering is portrayed by the activities and instructional strategies they choose.

RELATIONSHIPS BETWEEN S, T, E, AND M IN ELEMENTARY CLASSROOMS

As I mentioned above, working on real-world problems often means calling upon engineering, science, and mathematical knowledge and skills. Convincing a client or a safety review board that a bridge or medicine will function as expected requires explanations based in scientific fact and backed with mathematical calculations. Mathematics, science, and engineering thinking are intertwined and interdependent. In the real world, most engineers and scientists draw from the deep bank of cross-disciplinary knowledge they have accumulated through study and practice. However, in elementary schools, young students do not yet possess extensive knowledge, ideas, practices, or skills to guide their work—that is one of the purposes of schooling. Students at these ages are still actively assembling their S, T, E, and M toolkits. Thus, as a recent report about integrated STEM from the National Academy of Engineering recognized, "Connecting ideas across disciplines is challenging when students have little or no understanding of the relevant ideas in the individual disciplines" (Honey, Pearson, & Schweingruber, 2014, p. 5).

As a result, teachers are tasked with helping their pupils develop knowledge and skills in each of these individual disciplines. Current accountability systems, especially standardized tests, assess proficiency in each content area separately. In most schools, teachers need to consider how their instruction will affect these metrics. Therefore, accountability and the resulting emphasis of school-based instruction are important to remember when thinking about integrated STEM lessons and activities. At present, in many states, students and teachers are held to measures that are not well aligned with the important features of integrated STEM. As we think about connecting students' learning through integrated activities and projects, we must remember how existing assessments evaluate students and teachers. We must provide supports to help mitigate this gap as best we can (while also working to change assessments).

Our path to integration started with two commitments. First, we knew we wanted students to engage in problem-based engineering challenges because problem solving is fundamental to engineering. Engaging in relevant and meaningful problems can often motivate students and help them learn better. We recognized that any real-world problem would touch upon and draw from a number of other disciplines—not only science and mathematics, but also social studies and English language arts. Second, we recognized the challenge we faced in convincing educators to consider a new discipline. We knew there was very little appetite for additions to the school day, so we knew we would need to make a compelling case for elementary engineering by proving its value. We would need to show how students benefitted. We knew that to convince schools and districts to dedicate resources to engineering, we would need to gather evidence—in the form of large-scale, bubble-scan, statistical assessments—that proved students learned engineering concepts and skills from the engineering lessons. Our case for including engineering at the elementary level would be further strengthened if we could develop curricula that not only bolstered students' understanding of engineering but also of an assessed subject already in school curricula, perhaps science or mathematics. If we could demonstrate increased performance on standardized tests, all the better.

As my team and I began thinking about how to introduce engineering to elementary classrooms, we turned to our experts, our elementary teacher consultants, for advice about how to do so. They strongly suggested that we needed to integrate engineering with topics that they were already teaching. Their rationale was primarily practical—they were concerned with how they could realistically add another subject to their already-packed school day. As we considered this advice, we thought about which subjects engineering relied on most heavily and decided to integrate engineering primarily with a single subject, science. We recognized that we would need to carefully design lessons and activities that asked students to draw upon concepts they were learning in science and apply them in new ways as they engineered—thus potentially bolstering their understanding of science. We generated a list of the 20 most commonly taught science topics in elementary school: plants, magnets, weather, forces, and so on. These topics became the structural backbone for our curriculum. Each unit would focus on one science topic and we would design the engineering activities to integrate with and reinforce associated science concepts. We identified the science focus first. Then we determined the engineering focus.

Because engineers solve a vast array of problems, we narrowed the scope and created concrete situational problems for students to solve. The best plan, we decided, would be to have each unit focus on a different field of engineering. This idea stemmed from the research we conducted that showed that students associated engineering primarily with certain fields such as civil, electrical, or mechanical. To help students understand the wide range of work that engineers do, we developed units that feature many different fields of engineering, including chemical, green, biomedical, materials, geotechnical, environmental, and transportation engineering. As we began each unit, we specified the engineering field that would connect to the science topic and then used the library, Internet, and interviews with practicing engineers in that domain to distill a few core ideas of that specific field. For instance, acoustical engineers work to dampen or amplify sound. These foundational understandings became the backbone of our units.

Once we had a science topic and an engineering field, I challenged the team to develop activities that help teach kids about both. We brainstormed engineering activities that would require students to use science concepts as they worked as engineers to solve a challenge germane to the engineering field of focus. As we did so, we first laid out unit learning objectives for science—for example, "Students learn that sounds are produced by vibrations"—and for engineering: "Students learn that sounds can be damped in different ways, including stopping vibration at the source and stopping the transmission of vibration through matter" (Engineering is Elementary [EiE], 2011a, p. 55).

We used these learning objectives to guide the engineering activities we created. We also used them to structure the assessments we created to measure what students learned from the unit. We found that we could closely integrate science and engineering in ways that enabled us to measure student learning gains in both (see Chapter 7).

We chose to integrate engineering primarily with one discipline (science) instead of with all STEM disciplines. Why? Because we were aiming to develop a curriculum that could be used nationwide. In the United States, we do not have a national curriculum. Teachers across the country teach math and science topics in various grades and configurations. Our curriculum's close reliance on science meant that we wanted teachers to select their engineering units based on the science topics they were teaching. Across the country, students might learn about a particular science concept in any one of a number of grades. For example, students might learn about the structures and functions of plants in 1st, 3rd, or 5th grade, or in multiple grades. We devised a strategy to create the engineering curriculum so that units could be implemented in any of grades 1–5. Thus, we could closely integrate two disciplines: science and engineering. However, we could not know which mathematics concepts students had already studied or which concepts they would be in the process of learning when they engaged with a particular engineering unit. As a result, it was impossible for us to create lesson plans that also closely reinforced the mathematical ideas students were learning that month.

In contrast to science, we view mathematics as "connecting" rather than "integrating" with our engineering lessons. We believe the relevant mathematical knowledge must stem from the engineering or science activity, not from what the students are currently studying about the subject in school. We identify where meaningful mathematics is used during the engineering activity and we have created additional mathematics extension lessons that draw out or support these concepts. This means that the mathematics that students might be currently learning in their math class—for example, ratios—might not be the math they employ during the engineering project. If the topic of study happens to be relevant, teachers can make connections. But in many cases, trying to force a connection can lead to contrived, largely meaningless activities.

Instead, our engineering activities commonly require students to apply their measurement and graphing, averaging and rounding, and data analysis skills. These concepts pervade the Common Core mathematics standards and teachers appreciate that students apply these skills in meaningful ways. One 4th-grade teacher who had her class engineer parachutes described the opportunity that the activity offered her students for their mathematical literacy (you can watch this interview at eie.org/book/2a):

> The one thing about measuring: I think it's important for kids to measure a lot. It's really hard in elementary school to teach measurement, because they're not out there really physically doing measuring things. You do it here and you learn it there in conversions . . . , but really using that tool, that's time-consuming. This is an opportunity to use those tools to measure. . . . I think these [engineering] lessons, especially this one, gives them a great way to use measuring tools that they wouldn't have had otherwise.

English and social studies skills can also connect with engineering projects. Engineering in the real world takes place in a specific context and usually involves a client. Including these elements in school projects invites project-based learning and encourages intersections with other school subjects. For example, setting a context for a challenge can be done with a storybook or a newspaper article—providing opportunities to link to reading and English instruction. Asking students to think about how the technology might affect a community or inviting them convince a client that their design is worth adopting allows students to consider the societal implications of technology, cultural practices, and differing points of view, and provides practice communicating in written, graphical, and oral modes.

Our aim to create a curriculum that could be used nationally and accommodate the realities of U.S. schooling resulted in our decision to closely integrate three of the STEM disciplines— science, technology, and engineering—and to connect to the fourth: mathematics. But there is no singular "way" to introduce STEM into the classroom. In reality, the concepts interweave in various ways depending on the goals and projects. If you are a teacher or district creating curricula or activities and have specified topics and sequences of study for each of the disciplines, you might be able to create strong integration across all four topics.

In thinking about how to integrate S, T, E, and M in instruction, a couple of frameworks might be helpful. One considers current attempts at integration and which topics lend themselves to it. In their book, *STEM Lesson Essentials, Grades 3–8: Integrating Science, Technology, Engineering, and Mathematics*, Vasquez, Sneider, and Comer (2013) describe a continuum of STEM approaches to curriculum integration: disciplinary, multidisciplinary, interdisciplinary, and transdisciplinary. The authors point out that all of these approaches can have value and that teachers will likely use a mixture of them in their classrooms.

A National Academy of Engineering committee that reviewed scholarly research related to the design of integrated STEM instruction articulated three key implications that also merit careful consideration:

1. Integration should be made explicit. Most students do not spontaneously make the connections that are intended and thus need both "intentional and explicit support . . . to build knowledge and skills both within the discipline and across disciplines."
2. Students' knowledge in individual disciplines must be supported. It's difficult to connect ideas across disciplines, especially when students do not (yet) have a well-established understanding of the relevant ideas. They will need support to draw upon relevant math and science ideas as they solve engineering problems, to productively connect the ideas, and to create understandings that mirror those of normative science.
3. More integration is not necessarily better. There are both challenges and benefits associated with making connections across STEM subjects. These both need to be taken into account as instruction is planned. (Honey et al., 2014, p. 5)

No matter how you choose to structure STEM integration, an important benefit is that students are using knowledge and skills in purposeful and meaningful ways that contribute to overall understanding.

INTEGRATING SCIENCE, ENGINEERING, AND MATHEMATICS IN AN ELEMENTARY CLASSROOM

I close this chapter by sharing one example of an activity that integrates engineering, science, and mathematics to design a technology. The lesson is primarily an engineering lesson. We created it with the goal of introducing students to engineering and the engineering design process. As students work as green engineers to design a solar oven, we want them to use science concepts they have learned related to energy and heat transfer. In the lesson, students also need to use mathematics in authentic ways. This lesson is part of the Now You're Cooking: Designing Solar Ovens (EiE, 2011b) unit. Constructing solar ovens has been a popular activity in science classes for years. We decided to mix up the traditional activity that has students create an oven using aluminum foil and dark surfaces. Instead, we provide students with a standard base oven design and challenge them to design *insulation* for the solar oven, which allowed us to introduce concepts related to green engineering (see Figure 2.1). Students' designs are evaluated based on the ovens' ability to retain heat as well as how environmentally friendly their insulation is.

You can watch a 3rd-grade classroom in Marietta, Georgia, and a 4th-grade classroom in Washington, D.C., engage in this engineering challenge at eie.org/book/2b.

A storybook introduces students to the engineering design challenge. Prior to starting the design challenge, students also explore a tool used by engineers to evaluate the impact of a technology on the environment: a "life-cycle assessment." Students learn about this tool and conduct a simple life-cycle assessment of paper—a common product they interact with daily. They will use what they learned from this activity about how materials can impact the environment and from their study of energy in science class as they engineer their solar oven insulation.

The lesson begins by prompting students to think about the design of a solar oven. The teacher asks them what the purpose (or goal) of a solar oven is and they review its four parts: the reflector, oven box, window, and cooking pot (see Figure 2.2). With this in mind, she prompts students to reflect on why the sample solar oven is designed as it is, asking: "Why do you think the reflector is covered in aluminum foil?" and "Why is the hole in the top of the oven covered in clear plastic?" (EiE, 2011b, p. 87). Then the teacher invites the students to connect their thinking to energy, particularly heat energy. She

Figure 2.1. Designing Solar Oven Insulation

Figure 2.2. Parts of a Solar Oven

Parts of a Solar Oven

1. Outside of Solar Oven

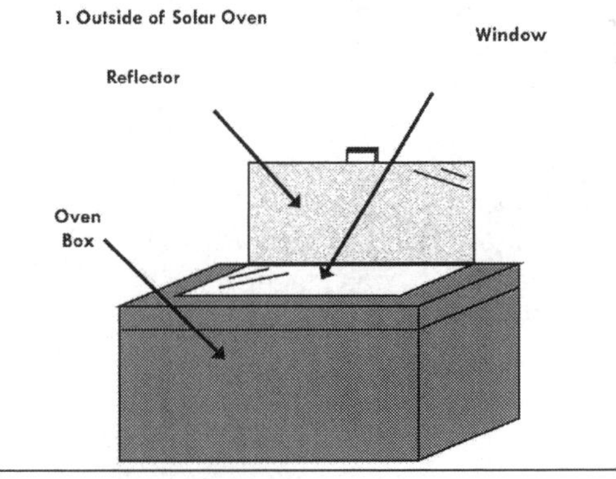

2. Inside of Solar Oven

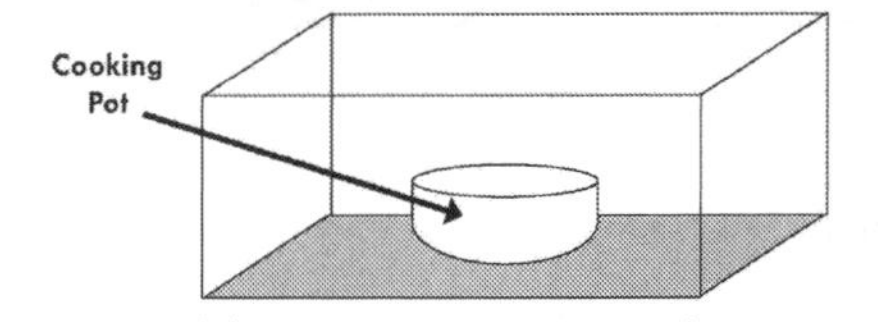

prompts students to consider the source of energy for the oven and how it works (the solar oven traps heat energy from the sun and transfers energy from the air to the food). She also helps students recognize the important need to trap the heat energy with various parts of the design. For example, the window at the top of the oven is covered by plastic to trap heat in the oven. Insulation can also trap heat.

In the first part of the design activity, students gather "scientific" data that can inform their engineering decisions and designs—they will conduct controlled experiments to test how the available materials (aluminum foil, foam, felt, plastic, and newspaper) perform as thermal insulators. First, the class reviews what insulators and conductors are. They identify properties of each of the materials and make predictions about its abilities as an insulator. And the class establishes a fair test with a control that they will use to measure how well each material traps heat. They will test each material twice—as a "flat" material and as a "shredded" material.

Students make predictions about which materials they think will be the best thermal insulators. Then they work in small groups to test 10 types of materials. Before they put each material in a cup, which will be placed in an ice bath, the teacher asks students what they think will happen to the air temperature in the cups and why. This prompts students to think about how the energy is flowing in this system—the heat energy is flowing out of the hotter cup into the colder ice bath. The students collect data by recording the temperature of

Figure 2.3. Collecting Temperature Data

the air in the cup every 30 seconds (see Figure 2.3). When data collection is complete, students analyze their data by calculating the change in the temperature for each material.

As a class, the students describe what temperature changes they observed and the teacher again asks students to think about the movement of heat in the system with science questions such as "Where did the heat energy go? Why do you think so?" (EiE, 2011b, p. 99).

The students continue to reflect as they analyze their data. The class creates an insulator scale from Poor Thermal Insulator to Good Thermal Insulator and assigns each test material (composition and flat/shredded) a place on the scale based on the temperature change they measured. After ranking the materials, students generate bar graphs comparing the temperature changes of the flat and shredded treatments for each of the five materials and compare them to the control setup that had no insulators.

With these data depictions, students are asked again to make observations and explanations for why the shredded materials insulate better than flat materials, in terms of heat energy and movement of air. The students reflect on what properties of materials work well as insulators before they move on to think about the environmental impact of each material.

In the last phase of their exploration of materials, students consider environmental impact. They identify whether each material is natural or processed and how much of it is needed (Can they reduce this?), whether it has been used previously for a different purpose (Is it being reused?), or whether they can recycle it (Can it be recycled?). They create a second scale that portrays the environmental impact and place each material on the scale from Least to Most Environmental Impact.

With guidance from the teacher, the students reflect upon the two scales they created (thermal insulator and environmental impact). They weigh tradeoffs between the materials and deliberate about which they might use in their designs.

As they gather more knowledge to apply to their design choices, students ask questions and conduct experiments that reinforce science concepts about energy. They also utilize mathematical tools that allow them to compare and understand their results. As students move on to brainstorm ideas (see Figure 2.4) for insulation design, they will (or they should) draw upon this knowledge.

The students then engineer the insulation for their solar oven, balancing thermal properties as well as environmental impacts. They plan a design and create it.

Students test their designs by placing the ovens in the sun or under a bright light and recording the temperature inside as it rises at 5-minute intervals for 30 minutes. At that point, they move their ovens to the shade and record the temperature inside the oven as it falls every minute for 10 minutes.

Figure 2.4. Sketch of Oven Insulation Design

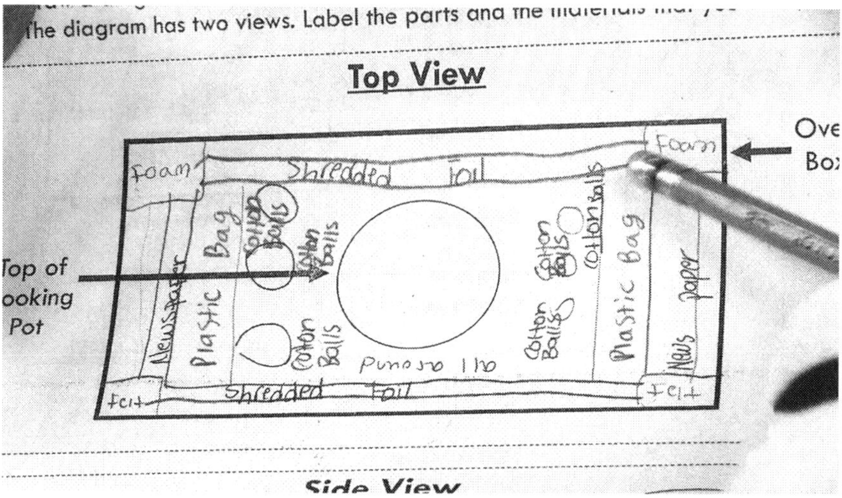

Students evaluate their designs with a:

- Heat Score (which compares the maximum temperature their design reached to a control);
- Time Score (how long their design holds its heat before returning to the original temperature when placed in the shade); and
- Impact Score (the environmental impact of the materials used).

Using these individual metrics, they calculate a Total Score. Based on their observations of their oven, its performance, and those of their classmates, each group thinks about what worked well about their design and how it might be improved. Then they redesign and test at least once more.

We designed the Now You're Cooking: Designing Solar Ovens unit so students could engage in engineering in authentic ways. They were introduced to a real-world context and then had the opportunity to develop knowledge relevant to the problem at hand as they honed their comprehension of heat transfer and insulators. This permitted them to draw upon such knowledge as they worked through the challenge. The solar oven insulation engineering challenge asks students to call upon and further develop science concepts related to energy, energy transfer, and insulators. Through application, students build a deeper understanding of science concepts that include the following:

- Some materials transfer heat energy more readily than other materials (there are thermal insulators and thermal conductors).
- Heat is one form of energy.
- Energy can be transferred from one object or material to another.

- Heat energy always moves from warmer locations to cooler locations. (EiE, 2011b, p. 78)

Our research demonstrates that students who engage in this integrated engineering unit do learn the science better than students who only study the science without any engineering. Not surprisingly, by manipulating science concepts in relevant and meaningful situations, students come to understand those concepts in a deeper way. Integrating science and engineering in thoughtful ways can bolster students' engineering and science understanding. We'll return to this claim in Chapters 5 and 7.

Additionally, our solar oven activity provides students with an opportunity to see how math is useful and important. The students collect data, create tables, and perform calculations as they work to maximize the functioning of their design with respect to the stated criteria. For many students, these mathematics topics might not be the ones they are currently studying in their math classes. However, numeracy skills, such as measurement, calculating differences and rates of change, and creating graphs to display and analyze data, need to be reinforced continually throughout schooling. Students have a vested interest in their engineering designs and their performance. They want to make them better! By using mathematical tools and techniques in meaningful ways, they can make sense of the data they collect—data that can help them improve subsequent designs.

The 4th-grade teacher in Washington, D.C., describes how her students made connections between engineering, science, mathematics, and other school subjects as they engaged in this unit:

> I have taught a number of different engineering curricula and then created my own units, but it's always based on the EiE model. Specifically with EiE, we have a real-life problem that is existing in our time somewhere in the world that people are confronting. Already with EiE, it allows kids to connect. It also allows me as a teacher to find other examples to connect. In this case with solar ovens, we were able to start with the story, but then also bring in real images and real stories from places in Botswana and other western African countries that are actually dealing with this. It makes kids feel like, "I'm solving a real problem that actually exists, not just some made-up thing." That makes EiE really unique. Those cultural and real problems that are existing right now for kids to solve. It makes them feel like they're really doing the work.
>
> Then there's the scientific component—that second lesson that allows kids to collect data, test materials, learn about a scientific concept so that they can apply it. Because we always tell them that engineers use their scientific knowledge as part of the process, but sometimes they don't have

> that scientific knowledge. So it allows that real "in" for a classroom teacher to dig deeper at specific scientific content.
>
> [In lesson] three, you're always looking at materials. Those lessons where you look at materials and what parts of those materials you might want to use and using the data to come up with solutions. Even as we did this unit, you could see this progression of understanding. They didn't have any scientific knowledge around materials at the beginning and thermal insulators. They had minimal [knowledge], after we talked about the life cycle, assessment. They gained more after testing, but when they really started to apply that is when they really put everything together.
>
> In lesson four . . . they're able to use the engineering design process to use their scientific knowledge they have received during this, to use the data from the math perspective. To use their own creativity, and actually test and then have time to improve, is really unique.

Like this teacher, as you select or design engineering activities, consider which of the STEM disciplines you will include and how you will highlight for students the differences and overlaps among them. Science, technology, engineering, and mathematics are clustered together in STEM because they often intersect in beneficial ways. Inviting students to use science and mathematics in authentic ways to solve engineering problems can help them understand how and why these subjects are relevant and interconnected.

Part II

CONSIDERATIONS FOR CURRICULUM DESIGN, INSTRUCTION, AND LEARNING

CHAPTER 3

Core Engineering Concepts

What Does Engineering Look Like in the Elementary Classroom?

After one of my first EiE professional development sessions, a 3rd-grade teacher confessed that she had almost skipped the workshop. The night before, she had been unable to sleep, anxiously trying to imagine how she would ever be able to teach engineering to her students. "If science is scary, engineering is terrifying," she asserted. She had no idea what engineering looked like with young kids and she had no background in the discipline itself. This was not the first time someone had shared these fears with me. I witnessed this initial trepidation often. The teacher went on to say she was glad she did attend, despite her anxiety: "Now that I understand what engineering looks like for children, I see how it can work in my classroom and how engineering will benefit my students. I can do this. They can do this!" When we first began developing the EiE curriculum, no one knew what elementary engineering instruction looked like, and it was the fear of the unknown that led many people to reject the very notion. However, I believed, and still do, that once educators have an understanding of what core engineering concepts are and how they can be taught to students of any age, they can begin to imagine it in their classrooms.

At this point, maybe you're wondering what engineering looks like at the elementary level. I'll admit that I didn't know either when I began. I had many questions: What were the most important engineering ideas to communicate? What could young students realistically do? What would their "doing" look like in a classroom setting? Theoretically, it made sense to build students' problem-solving abilities. But there were many practical details we needed to understand better before we could offer advice about how to introduce engineering to students. Seeking answers to these questions, and many more, my team and I got to work. We read many, many books and articles. We talked to practicing engineers to distill what they considered core engineering ideas (for a more complete review of these, see Chapter 4). We reviewed psychological and educational research focused on children's development and elementary science learning. We observed children at play and students in classrooms. We consulted many teachers and parents of elementary-age students.

Through these efforts, we identified two core engineering concepts:

- Engineers use a process to design technologies.
- Materials have properties that affect how they can be used.

These ideas are fundamental to the discipline and its practice. Though we did not know what elementary engineering instruction would look like in the classroom, we knew our curriculum had to develop students' understanding of these two foundational concepts. In the first part of this chapter, I explore each of these framing ideas and consider how elementary students can engage with these concepts in age-appropriate ways. Then I use a vignette to bring these to life, describing how these ideas manifest in a 1st-grade classroom.

CORE ENGINEERING CONCEPT: ENGINEERS USE AN ENGINEERING DESIGN PROCESS

Perhaps the idea most central to engineering is that engineers use a systematic, iterative process—the engineering design process (EDP). This process guides them as they work to solve a wide variety of problems in many different fields such as civil, biomedical, environmental, and chemical engineering.

Flashes of insight and tinkering contribute to product development. But before a technology is released as an "acceptable" product, engineers must design and test their products numerous times, along the way ensuring that important stakeholders approve their design and process. Buildings, bridges, medicines, and even hair dryers, for example, must meet industry requirements and regulations to protect consumers' safety. Engineering firms and companies need to document their work and results and justify their decisions—this is one reason why most of them employ methodical processes to develop their technologies. For instance, the Dyson company worked for 4 years to design its innovative hair dryer. Their engineers examined hair from people of different races with a scanning electron microscope to understand the differences and then studied how various types of hair behaved in turbulent air flow. Did it tangle? How and why? Engineers who worked on the project modified bladeless fans and digital motors that the company had developed for previous technologies, incorporated them, and tested them in the hair dryer. They collected thermal images to understand how their dryer models dispensed heat over strands of hair, over time. (View images of prototypes they developed at eie.org/book/3a.) The dryer uses heat sensors that take the temperature 20 times per second, which allows it to regulate temperature to protect both hair and the person using the device. As part of the engineering process, all the components and variables needed to be tested and optimized. By the time the new hair dryer was released, the company had spent $71 million on development, engineers had tested 1,010 miles of human hair, and more than 100 patents were pending from Dyson's innovative approach (Rhodes, 2016).

Although this example comes from corporate engineering, its lessons are applicable across fields and settings. Professional engineering projects have dozens and dozens of steps. Professional engineers might work on just one or two of these steps, however, and then pass the information, sketches, models, or findings to another team to build upon. For example, an engineer might receive a diagram of a new possible airbag from her colleagues. Her job as a material engineer might be to run calculations and models to make recommendations about which materials could be used for the inflating bag that meet the required specifications. Having done that, she might pass her team's recommendations on to another team that constructs a prototype and conducts tests on it to collect physical data. Ultimately, engineering requires and is defined by careful research, problem definition, testing, and analysis. Engineers optimize their solutions within a set of constraints. They learn to support their ideas with methodical, documented processes. In some instances, such as the hair dryer, the careful work results in a product that is innovative and marketable. In other cases, engineers need to use the product design to convince their clients it can function appropriately and safely before it is ever even created. For example, no town or state will invest $100 million to construct a bridge, no matter how innovative or beautiful, without knowing that a complex set of mathematical calculations and extensive testing undergird its construction and show that it will support the anticipated load.

Methodical and data-driven approaches to solving problems are not necessarily characteristic of children's (or adults') design work. Novice engineers often dive in and start building without a plan (Hill & Anning, 2001). Halfway through the project, they may realize the various pieces don't fit together or that they have run out of supplies. I remember watching my 4-year-old niece build a fairy castle. She had a vision for what kind of castle she needed and she dove straight into its construction. She first used all of her longest blocks to build the lower parts of the walls. After she ran out of long blocks, she started to use shorter blocks to build the upper parts of the walls. By the time she was ready to build the castle's roof, she realized she had a problem. The remaining blocks could not span the structure. They were too short! Because she did not plan how she would allocate her resources, she could not construct the type of roof she had originally envisioned. We discussed how she could have designed her castle differently, and she recognized that she should have "saved" the long blocks to make the roof and used the shorter ones for the walls. Of course, she was not inclined to start all over again. She decided instead that her fairies would live *en plein air*. Although this decision was fine for her imaginary play, it isn't a helpful tactic in the real world, which is why we need to scaffold children's planning abilities. As we work to develop students' engineering sensibilities and problem-solving abilities, we need to help them move from an approach in which they create technologies on the fly to one that is more planned and systematic—while still having fun, of course.

Creating an Elementary Engineering Design Process

Budding K–12 engineers need to become familiar with the cyclical process of innovation that underlies engineering. Therefore, most educational programs use an engineering design process (EDP). There are many, many variations of EDPs. When I started, there were industry, university, high school, and middle school processes. I knew these were too complex for elementary students. But an age-appropriate elementary design process did not yet exist . . . so we would need to create one. To do so, I knew we had to talk with teachers—after all, who knows what works in an elementary classroom better than an elementary teacher? I convened an advisory group of classroom teachers. I called school principals in my region of Massachusetts that served high numbers of students from the underrepresented and underserved communities I wanted to reach. I explained that I was hoping to develop some supports that could help elementary teachers implement the new Massachusetts standards, which included engineering for the first time. I asked each principal to suggest two teachers who might serve as consultants to this project. My only requirement: that the teachers were not afraid to teach STEM subjects. After a number of calls, I had collected the names of eight curious elementary teachers from four local districts that served diverse students. This advisory group of close collaborators and pilot testers, which grew in number over time, became invaluable to our curriculum development team.

To create an elementary EDP, we first needed to help teachers develop a working sense of what engineering was. Teachers engaged in a few simple design challenges that my team was developing: They built walls, bridges, water filters, and windmills. They recognized that they used a similar process to engineer all the solutions. We discussed how practicing engineers also use an EDP. The consulting teachers agreed that a simple EDP could serve as an "organizer" and guide students' work at the elementary level. We showed them the eight-step EDP in the Massachusetts Science and Technology/Engineering Framework (Massachusetts Department of Education, 2001). They immediately rejected it as too complicated for their students. So, we brainstormed possible models with varying numbers of steps. The elementary classroom experts emphatically insisted that whatever process we came up with could only have five steps—like the fingers on one hand—or their students would not remember it.

Distilling a process to only five steps required careful, disciplined thought. Our observations of young children engineering suggested that children loved constructing and testing their designs. That's the fun part! That part of the process would easily occur without including it in the EDP. Less obvious to students and teachers were the steps that preceded and followed the construction. We decided that students' designs and understanding of engineering would be strengthened if we could call attention to the steps that prompt them to reflect on their goals, the criteria for a successful design, and what they could bring to the problem. We also wanted to encourage students to generate multiple possible solutions before homing in on one to pursue. Based on the teachers' suggestion, we developed our five-step process: Ask, Imagine, Plan, Create, Improve (see Figure 3.1).

Engineers start with a goal—something they are trying to accomplish. Then they move through a process asking these sorts of questions along the way:

ASK:	What is the problem?
	How have others approached it?
	What are your constraints?
IMAGINE:	What are some solutions?
	Brainstorm ideas.
	Choose the best one.
PLAN:	Draw a diagram.
	Make lists of materials you will need.
CREATE:	Follow your plan and create something.
	Test it out!
IMPROVE:	What works?
	What doesn't?
	What could work better?
	Modify your design to make it better.
	Test it out! (EiE, n.d.)

The engineering design process can be tailored for the age and experience of any engineer. Students need scaffolds that are age-appropriate and reflective of their developmental capabilities. Students may begin engineering as preschoolers. Three-year-olds might experience engineering problem solving as three simplified stages—preparing to design (Explore), creating and testing the design (Create), and reflecting/improving on the design (Improve)—because children at this age can manage three steps (see Figure 3.1). The five steps of the elementary process build upon this basic EDP. And as students move from elementary school to middle school, they become capable of distinguishing more detailed steps, so the EDP can become more complex. Our middle school EDP has eight steps: Identify, Investigate, Imagine, Plan, Create, Test, Improve, and Communicate (see Figure 3.1). The most important thing about using an EDP is to get kids engineering reflectively—not just doing crafts or recipes, but working systematically, evaluating their designs, and improving them to accomplish a goal.

Figure 3.1. Preschool, Elementary, Middle School EDP

Preschool EDP - Elementary School EDP - Middle School EDP

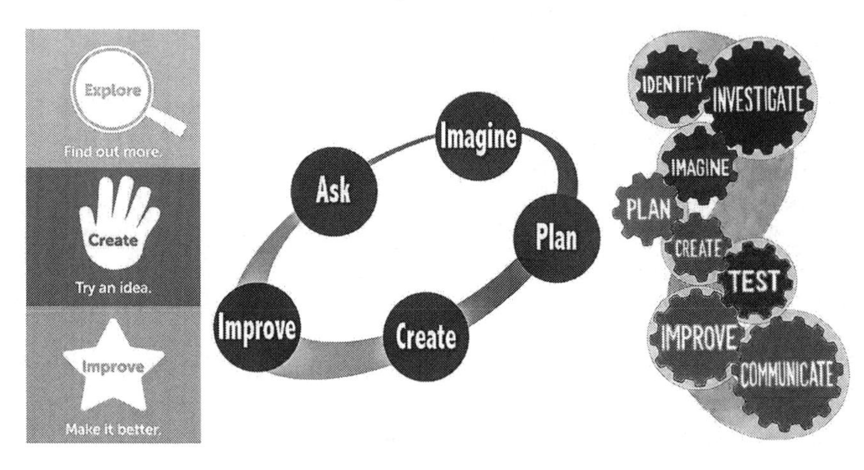

Important Characteristics of the Engineering Design Process

It is vital to note that any EDP is cyclical and iterative—there's no official start or end point. You can begin at any step, focus on just one step, move back and forth between steps, or repeat the cycle. In the real world, most engineering projects actually begin with the Improve step—someone thinks of a way to make something better or identifies something that needs fixing. Technologies are constantly evolving: The smartphone evolved from the cellphone, which was inspired by a portable phone, which came from a push-button phone, which grew from a rotary phone, which built on a candlestick phone, which came from a hand-crank wall phone. Similarly, there is no end point for the engineering design process; technologies can always be refined. A cycle begins again when an engineer starts asking questions about how to improve a technology.

The engineering design process is not meant to be a set of rigid steps that students memorize. Rather, it is intended to be a tool to organize and structure work—it's particularly helpful for guiding students' work. In reality, engineers, including student engineers, move back and forth between steps. For example, as they put together a plan for the design of their technology, students might raise a question about how a particular material behaves. This could prompt them to conduct research via a short, hands-on investigation so they can apply their findings to their planning process for their design. Similarly, as students begin to sketch a plan for a package for a plant that must keep it alive, they might wonder whether cotton balls or pieces of sponge absorb more water per ounce. A quick experiment will help them determine which is the more absorbent material and could inform their design. The EDP and its steps can orient students to the goal of the task for that day, as well as how it fits into the overall goal of the design challenge.

Engineering Design Process as a Problem-Solving Tool

My work with K–12 teachers has highlighted the benefits of introducing students to the EDP as a structured problem-solving process for engineering activities. A number of teachers have also used variations of it for different subjects. Over the past dozen years, teachers and schools using our five-step EPD have shared how they use this method as a unifying process across their curriculum and beyond. For instance, many teachers use a "writing process" to teach students to write that includes steps something like Ideas & Plan, Draft, Revise, Edit, Publish, and a "problem-solving process" during math lessons that generally includes steps such as Understand, Plan, Do/Solve, Check.

Educators also utilize the EDP to make carefully reasoned and data-driven decisions. For example, one 4th-grade class used it to figure out where they would take a fieldtrip. They decided that they would solve this problem and structure their decision using the EDP. Together, the students generated questions, criteria, and constraints. They brainstormed possibilities, identified variables, weighed them in importance, and assessed each option for each of the variables. Then they analyzed the data to make a recommendation. A different elementary school in North Carolina that has adopted an engineering theme uses the EDP schoolwide for many things, including student discipline. When a student arrives at the principal's office with a disciplinary referral, the student and administrator work through the EDP with the goal of identifying other ways the situation could have been handled. The student asks what the problem is, imagines some other ways for handling the conflict and some consequences for this infraction, and puts together a plan that will guide future behavior.

Our primary aim has always been to instill within students a love for solving problems. Engaging students in the EDP helps them build confidence in their abilities to approach unknown problems and work through them thoughtfully, making decisions based on evidence. It fosters strategies for problem solving, challenges them to innovate and think out of the box, and encourages them to persist when an initial idea or design does not work as hoped.

CORE ENGINEERING CONCEPT: MATERIALS AND THEIR PROPERTIES

> We needed to know the properties and the [testing] helped me out a lot . . . because then I figured out which [material] was stickier, which was stronger, and then it was easier to know which to use in the [design].
>
> –4th-grader

As this 4th-grade girl notes, materials matter: In fact, the second core idea that permeates engineering is the importance of materials and their properties. The materials that are used in engineering solutions—in both the real world and the classroom—all have their own properties. They might be fuzzy, transparent, strong, lightweight, inexpensive, malleable, and so forth. These properties dictate how the materials behave and should drive which ones are selected for use in a technology. All engineers need to consider materials, including their behavior under certain conditions, their interaction with one another and their environment, their cost, and their aesthetics, as they design technologies.

As we engage with our world, we build our knowledge of materials. Adults have a much more developed sense of materials and how they can be used than young children do. For example, if you ask an adult which material will best pick up pollen—marbles, erasers, pom poms, tin foil, or pipe cleaners—most will recognize that fuzzy materials, such as the pom-poms and pipe cleaners, are the best choices. Adults understand not only how an individual material might behave, but also are able to generalize about the properties of materials; for example, fuzzy things are usually pretty good at picking up small particles. Thus, adults are able to extrapolate and make conjectures about how unfamiliar materials might behave based on their properties.

The general characteristics of materials are not as obvious to children. Infants are not born with such knowledge; children develop it over time through interactions with and manipulation of the world around them. Students need rich opportunities to explore materials. They need to touch, try, and talk about materials and their properties. As students start to think about the problem they are going to solve, they should gather information and think about the dimensions of the problem. This "problem definition" or "problem scoping" can draw students' attention to important attributes of a successful design. For example, a class has to design hand pollinators. These hand pollinators need to pick up and drop off pollen (modeled using baking soda) for a model flower. Through a guided conversation, students need to identify what a pollinator should be able to do. For example, to be successful, a hand pollinator needs to pick up pollen and drop off pollen. Generating criteria for the technology helps spark a discussion about the materials and their properties.

Students are actively building their understandings of materials and their related vocabulary in elementary school. To ensure that students have equitable access to engineering and to ground their conversation in real experiences, elementary students should get a chance to systematically explore the properties of different materials that are available *before* they are asked to design and build with them. Thus, part of the Ask step is dedicated to helping students learn more about the materials they will use. Each group receives a set of material samples, and touches and manipulates them to become familiar with them. The class works together to create a "Materials and Their Properties" chart (see Figure 3.2).

This allows students to have a sensory motor experience, construct more specific understandings of how they might use the material, and build more precise and expanded vocabularies as they describe what they observe.

Figure 3.2. Materials and Their Properties Chart

Materials and their Properties

Material	Properties
pompom	fluffy, soft, round, red (and other colors)
tape	sticky, flexible, transparent (clear, see through)
aluminum foil	light, shiny, folds, silver reflective
pipe cleaner	flexible, can shape it, smooth, hard wire, stiff
marble	heavy, hard, smooth
eraser	pink, smooth rectangular prism

Students can then make predictions about which properties might matter in the construction of their designs and systematically test the materials to see how they perform. As they compare various materials and their properties and observe how they function, students can build their knowledge of the world around them. They can then select appropriate materials for their design considerations. Instead of choosing a material because it is pink or metal—two favorite material properties of young students regardless of the challenge—students make decisions based on how the materials meet the larger design criteria.

Children's social, cognitive, language, physical, and emotional development all need to be considered to choose and implement age-appropriate engineering experiences. We have found that the younger the child, the more challenging it can be to develop an engineering challenge. Younger students know less about science and have fewer math skills to draw upon. They are still developing fine motor skills, logical reasoning, and teamwork. They need simple explanations, fewer variables, and more supports as they work through a problem. These developmental constraints prompt careful reflection about the fundamental engineering concepts and competencies that educators hope to nurture.

The engineering design process and materials and their properties are two core ideas in engineering. What does it look like as elementary students learn about these in a classroom? In the next section, we'll peek inside a classroom to see what transpires as students explore the properties of materials and apply the engineering design process. We'll follow Ms. Mock of Lake Elmo, Minnesota, as she teaches an engineering unit called The Best of Bugs: Designing Hand Pollinators (EiE, 2011c), about plant pollination and agricultural engineering. You can view short videos of these lessons at eie.org/book/3b.

CORE ENGINEERING CONCEPTS IN THE CLASSROOM: 1ST-GRADERS DESIGN HAND POLLINATORS

Ms. Mock's 1st-graders gather in a circle for storytime. Ms. Mock reads the storybook *Mariana Becomes a Butterfly* aloud (EiE, 2008). At the front of the classroom, she displays the storybook illustrations on a wide screen. The storybook sets the context for the problem that the students will solve and introduces them to the engineering design process. In the book, Mariana's tropical plant no longer produces berries. Through observation, she realizes that her plant lacks a natural pollinator. She uses the engineering design process (EDP) to engineer a hand pollinator to pollinate her plant.

Ask: Identifying the Problem and Relevant Knowledge

Because young students benefit from repeated interaction with an organizing process or principle, Ms. Mock pauses frequently to ask her students, "What did Mariana just do? What step of the engineering design process do you think she's working on now?" She also stops periodically to pose questions that connect the story to her pupils' lives and knowledge. For example, she solicits her students' prior knowledge, asking, "What is pollen?," which prompts responses that associate pollen with flowers, bees, honey, and "flower dust." When she finishes the book, she asks the students to identify the challenge and envision themselves as engineers, saying, "Could you design your own pollinator if you had a plant that wasn't being pollinated? She [Mariana] gave you a couple ideas. Raise your hand if you are thinking of a way you could make a pollinator." Energized by the task and the possibilities, almost every student's hand shoots up. They are ready to use what they learned during their recent study of insect and plant life cycles to work as agricultural engineers.

With a partner, the students complete a worksheet that identifies the actions of the book characters and asks students to associate each with a step of the EDP. Experiencing the EDP in a narrative context first provides an approachable model and a concrete, shared referent for the class. Ms. Mock highlighted this feature of the book in an interview, saying, "They are 1st-graders. They love stories. It gave them a feeling of importance for what we were about to do. We referred back many times throughout the different lessons to Mariana and her problem and how she solved it."

Ms. Mock launches the challenge by asking her students to recall the problem and goal as presented in the storybook:

> ***Ms. Mock:*** When we read our story, what happened with Mariana? There was a problem. Olivia?
>
> ***Olivia:*** Her plant wasn't pollinating.
>
> ***Ms. Mock:*** Her plant wasn't pollinating. How did she solve that problem? Lucas?
>
> ***Lucas:*** With a parent helping out with an agricultural engineer.
>
> ***Ms. Mock:*** An agricultural engineer helped? What happened then? What did they have to do? Olivia?
>
> ***Olivia:*** They had to pollinate the plant.

Ms. Mock and her students begin the engineering design process by defining the problem. They ask questions about their design challenge. Then they figure out the properties of a good hand pollinator. Ms. Mock poses the question: "What must a good hand pollinator do?" They decide as a class that the pollinators they design should both pick up and drop off pollen. Ms. Mock opens a discussion about the materials exploration by asking her students to think about a number of questions:

> I think we should make pollinators. But we have to first think: What materials and properties of materials would work best for picking up pollen and dropping it off? If I gave you a bag like this [holds up a bag with samples of available materials], these are going to be the only supplies you have. In your head, I want to think about: How could I use what Ms. Mock gave me? How could I use [the materials in] this bag to create a hand pollinator?

Ask: Investigating What Materials' Properties Matter

Young students benefit from hands-on exploration (French & Woodring, 2014). Giving them a "materials bag" with a small sample of the relevant materials allows them to feel and manipulate the materials while the small size of the sample keeps students focused on thinking about their properties (materials cannot be manipulated in too many interesting ways at this size!). For this challenge, students can use marbles, tape, erasers, aluminum foil, pom-poms, and pipe cleaners. Ms. Mock instructs them to "write what property it has that will help it, or allow it to be a good pollinator" on a worksheet. With their partner, students eagerly open their materials bag, spill out the contents, and excitedly handle each sample. After the initial exploration, they examine each material in turn. As they do so, Ms. Mock circles the room. She asks questions and keeps the students on task. She asks one group if their pom-pom is "sticky." The students reply that it is not. Then she queries, "What other properties does [the pom-pom] have?" The students respond that the pom-pom is fluffy and rough.

As they work in pairs, students draw from their concrete experiences to construct their understandings of the materials and the words that can be used to describe them. A child in a different group exploring the marble initially states, "You can't see through [the marble]." Her partner disagrees. He holds the marble up to his eye, and announces, "Yeah, you can." Then they converse about whether or not a cotton ball is clear as they jointly build an understanding of the word *clear*.

Elementary-age students are actively building vocabularies, and they benefit from teachers guiding their observations as they develop a shared vocabulary about the materials. Ms. Mock convenes the class. Because these are younger students, instead of creating a "Materials" table with the students (as in Figure 3.2), she generates a set of opposite properties for them to reflect upon—rigid or flexible, heavy or light, clear or opaque. She asks her students which of these properties they think are important to consider as they choose materials for their hand pollinator.

As they consider each dyad, the students predict which properties will matter. The whole-class discussion ensures that students build their vocabulary as well as a useful knowledge base from which they can draw when they design their solutions.

After addressing and identifying properties, students next consider which materials will work well for the challenge's criteria:

Ms. Mock: We have to decide which materials will be good hand pollinators. You're going to test to see which material is going to be the best to use for a hand pollinator. So, I'm going to start with . . . I want to start with a marble. Yeah, this one says marble, see it? If I put this on the pollen, if it's a good pollinator, the marble will what?

Josh: Stick the pollen on it.

Ms. Mock: It will stick the pollen on the marble and then it it's a good pollinator, it will?

David: Work. Work. Drop it off.

Ms. Mock: Drop it off where?

Michael: To another flower.

Ms. Mock: To another flower, okay. Insects just land on pollen and they come over here and they go tap, tap, tap. Oh, I see a little. Don't blow! I see a little right there. But you have to also do the pom-pom, and the foil, and the pipe cleaner. You are comparing them to see which is going to be the best item or material to make your hand pollinator.

Ms. Mock demonstrates the procedure students will use. The students will dip each material in baking soda, which represents the flower's pollen, then gently tap the material against a cup three times to dislodge the pollen. She calls this the three-tap test. The white "pollen" falls onto black paper "flowers" so students can easily observe whether or not pollen is being deposited (see Figure 3.3). Because the students are still developing their fine motor skills, Ms. Mock will set up some of the more complicated components for them, such as allocating a small amount of baking soda to each testing setup.

Working in pairs, students test, deliberate, and record their data about how well each individual material picks up and drops off pollen on a worksheet. Because these are primary students, the data sheet asks them to circle how much the material dropped off: "No pollen," "Some pollen," or "A lot of pollen." To answer, they need to come to consensus about what they observe:

Olivia: I think it picked up some, right?
Alex: It just picked up some, but not a lot.
Lucas: Guys, I think this one and this one are tied [the same].
Paul: Yeah. That one looks better. This one looks better.
Lucas: No, they are both kind of the same.

The students do not necessarily adhere to the standard procedure. Some dip their materials into the pollen more than once and many groups tap more than three times. This can happen sometimes while students test in their individual groups. However, in public testing (later in the activity), the teacher continues to refer to the standard procedure, praising students who tap only three times. She works with the students to help them develop an understanding of why fair, standardized tests are so important.

After testing, Ms. Mock calls the students together to share what they have found, asking them: "What did you use? What did you do? What did you find?" The students, eager to contribute to the conversation, often talk over one another and finish one another's thoughts:

Lucas: The tape did most of mine.
Ms. Mock: What did it do?
Lucas: It took a lot of pollen and then spreaded [sic] it to the other flowers.
Ms. Mock: Did it drop it off? Noah, which one worked best for you or worked pretty well for you?

Noah: The pom-pom and the pipe cleaner.

Ms. Mock: Why do you think—this is for everyone—why do you think he's saying the pom-pom and the pipe cleaner worked the best? Predict why. Why do you think he said those two? What properties do a pom-pom and a pipe cleaner have that make this work so well for Noah?

Scott: It's fuzzy.

Ms. Mock: Ooh, did you hear him? Say that again.

Scott: It's fuzzy.

Paul: Cuz bees are like that. And they can collect the, like, they do the same thing like the pom-pom.

Ms. Mock: So the pom-pom did what?

Paul: It [pollen] sticked [sic] on the pom-pom.

Ms. Mock: Because of the . . . ?

Paul: Because of the fur.

Andre: It picks up things. Because hair is like string. It can pick up things on one of its sides.

Ms. Mock: Wow. You are doing some really big thinking today.

Figure 3.3. Three-Tap Test

Through such whole-class reflective sessions, Ms. Mock draws out the connections between the tests' results and the properties of the materials so that all students can learn and benefit from the class's insights and the conclusions. She reinforces the important takeaway messages that students should use in subsequent activities. She concludes the class by situating students' work within the larger EDP: "Did we ask a lot of questions? We were in the Ask phase. You asked me questions. I asked you questions. We tried things out because we were asking."

Imagine: Brainstorming Possible Pollinator Designs

The next day, students continue with the EDP. Ms. Mock revisits the five-step process and reminds the students what they learned the previous day. She asks them to identify which step they will tackle that day. They will Imagine, or brainstorm, original solutions for their hand pollinator. Ms. Mock explicitly states the students' task for this step: "I want you to brainstorm ideas for your design and I want you to list your ideas or draw a sketch or a picture. And then together you'll draw a circle around the one you think is . . . the winner."

Initially, her students work independently, sketching two to four possible solutions. The drawings are simple but students apply themselves diligently to capture the essence of their designs on paper. The youngsters are good at coming up with many unique, innovative solutions. They label parts and materials, sounding out the words *pom pom* and *pipe cleaner*. Representing ideas in sketches allows students to capture their thoughts so they can return later to further evaluate them.

Plan: Generating a Shared Pollinator Design

After brainstorming individually, students share their ideas with their partners. Together, they choose one idea or combine the best parts of their various sketches into one design, planning more details as they brainstorm. What materials will the hand pollinator be made from? What will it look like?

Eventually, each group draws another diagram or model of their new, collaborative design and, with prompting from Ms. Mock, labels the materials it uses. As they do so, they explain their design thinking to each other, put forth explanations for how it might work, and describe how they might build it:

> ***Andre:*** Yeah, but how are we going to put the pom-pom on?
> ***Karina:*** The pom pom?
> ***Andre:*** Put it on top.

These interactions require students to construct explanations about why they think design elements are particularly strong. It also requires them to work cooperatively with fellow students:

Karina: I have an idea. We could use this one [sketched design idea] and then just bend it [points to a part of the design].
Andre: Yeah. Okay.
Karina: So, let's say number three, but we'll just bend number three. How about that?
Andre: So, it would be half of my . . . it would look like your idea, but it will do the things that are in my idea.
Karina: Mnmhmmm.
Andre: So it's both of ours mixed up.

Ms. Mock nurtures such interactions—1st-graders are still learning to attend to the ideas of others. Her comments focus on teamwork, encourage compromise, and praise cooperation. She asks one group: "Do you guys agree on this plan? Do you like it together?" When the partners both agree, she affirms, "Ohhh, great teamwork." The desire to make a good hand pollinator motivates students to consider one another's ideas and come to consensus in an authentic context.

Create: Constructing and Testing the Pollinator

All this thinking and preparing sets the stage for students to dive into the Create stage with confidence and enthusiasm. This is the exciting part! Students will construct what they have drawn and test how it functions. Primary students are still developing their fine motor skills. Attaching components with tape, cutting materials so they fit with the flower model, and figuring out how the top and handle of the hand pollinator will operate together all require concentration. But the students diligently persevere, even when successful construction requires multiple attempts. The students are proud of their pollinators and eager to share their unique solutions with other classmates and their teacher (see Figure 3.4.) One student, Karina, exclaims, "Look what we've created so far. We used the foil, and we used the pipe cleaner."

It can be difficult for young students to adhere to their plans as they get access to piles of materials. To encourage the discipline students need, Ms. Mock compliments groups that are following their specific plans: "I like the way you are thinking. You are sticking to your plan. If you do that, you're sticking with this plan."

Once each group has constructed a technology, Ms. Mock calls the class together. They gather around a large circular testing table. Ms. Mock explains that it is time to find out: "Will your hand pollinator work?" She reminds them that they will test their design with the three-tap test. Because her class is young, she reminds students again how to conduct this test. "Three taps and lift. Is there any pollen?" One by one, each group tests its design publicly. The other students lean forward in anticipation of the outcome.

Figure 3.4. Hand Pollinator Designs

Ms. Mock scaffolds each test. She guides students through the testing procedure: "Here we go, testing. One, two, three. Good job. Three-tap test. Okay. Lift up your flower." And then she asks the designers to evaluate the results: "Did it work?" One group gets "a little" pollen to drop. The next group's design transfers "a lot" of pollen. In such cases, the whole class celebrates:

Class: [seeing the results] Whoooaa!
Ms. Mock: Did it work?
Testers: Yes!
Ms. Mock: Well, congratulations! Give them a hand! [students high-five one another]

With some frequency, a design does not work. This could be a time of great disappointment, tears, and distress, especially with young students. However, Ms. Mock treats such instances by referring students back to the EDP. The dialogue becomes almost a class chant:

Ms. Mock: Would you say your hand pollinator worked this time?
Testers: No.

Ms. Mock: What do you need to do?
Testers: Improoooove.

Testing publicly allows the teachers to help students perform standardized assessment of the technologies. But it serves an even more important function—all students get to see, learn from, and celebrate their classmates' designs. The degree to which these young students are interested in their peers' solutions is remarkable. It's also noteworthy that when designs do not work as expected, the kids are more eager to get back to work. As Ms. Mock reflects:

> When kids are engineering they are excited. These kids could not get enough of the hands-on activities. And when they created a hand pollinator and it didn't work . . . they automatically, instinctively wanted to make it better.

Improve: Making the Pollinator Design Better

Ms. Mock launches the Improve stage, saying, "You saw what your hand pollinator did. Take 3 minutes to make it better. How can you make your hand pollinator better? . . . Be thoughtful: How can I really get this hand pollinator to work?" Back in their pairs, the students generate lots of ideas, including:

> Make it longer.
> Use the pom-pom.
> Use more stuff to get more pollen.
> Get some tape. Tape it together.
> Need to tie it.

Taken as a whole, these types of comments demonstrate that students are attending to variables that matter to the hand pollinator's function as they think about improvements. Pollinators need to be long enough to get into the model flower (one shortcoming of some designs), they need to use materials that can transfer the pollen (such as pom-poms), and they need to be constructed in ways that improve the stability of the components (taping and tying).

Ms. Mock concludes her students' first engineering experience by praising their wonderful work and leading a conversation about which step of the engineering design process was their favorite and why.

In this classroom, we see primary students engaging with core engineering concepts at an age-appropriate level. Students tackle an engineering problem and anchor their work in the engineering design process. As part of this cycle, they explore materials and their properties to determine which are relevant to the challenge at hand. They conduct investigations to build a base of knowledge, and they refer back to this knowledge when they imagine and create their designs.

Young students require simple, concrete experiences. They are still developing their construction abilities, their understanding of fair tests, and their abilities to observe and explain. As students age and gain experience, they are able to:

- make more accurate predictions and entertain more hypothetical situations;
- transfer abstract knowledge to new situations;
- generate more accurate and sophisticated descriptions and explanations;
- reflect upon the findings of their peers, relate them to their own work, and use what they learn through such analyses to inform subsequent actions;
- balance tradeoffs and weigh how multiple components might interact;
- adhere to more standardized methods and procedures and recognize discrepant and outlying data; and
- engage in more detailed analyses.

Obviously, the complexity of the engineering challenge must be tailored to the audience. But all students can engage in engineering thinking and design and use a process to orient themselves to the goal of the work. (For more about engineering trajectories, view an engineering trajectories table at eie.org/book/3c, or see descriptions in Cunningham, Lachapelle, & Davis, in press, and Lachapelle, Cunningham, & Davis, 2017.) This chapter focused on two core concepts of engineering. The next chapter explores additional engineering practices and habits of mind in more depth.

CHAPTER 4

Engaging in Authentic Engineering Work

When I began working in K–12 engineering education, I hoped, but never expected, that engineering would become a mainstream part of state and national standards. Advocates, committees, curricular programs, and innovative classroom teachers who demonstrated what engineering looked like with students helped fuel the conversation that led to the inclusion of engineering in national and state standards. The Museum of Science led many of these efforts.

The Next Generation Science Standards (NGSS) are the third set of national standards (after Common Core math and English language arts) that many teachers need to learn and adopt. This can be overwhelming, especially for elementary teachers who teach all subjects and thus should become familiar with all three sets of standards and how those standards shape their classroom instruction. When I speak with classroom teachers, I find that only some are aware of the NGSS. They know that they will (eventually) need to understand and use the standards document, but they find it lengthy, complex, and confusing.

In this chapter, I first visit the Next Generation Science Standards to help make this task easier for teachers and to unpack engineering's core ideas and practices. Engineering is a new discipline for K–12, and its treatment in these standards needs work. Science and engineering are treated differently, which can make it even more baffling, especially because most K–12 teachers are new to the latter discipline. As researchers give more attention to classroom engineering and collect more data to support the effectiveness of certain engineering approaches, classroom engineering and engineering standards will continue to evolve. For now, I discuss the NGSS components and practices through an engineering lens and show how they can play out in classroom instruction given our current understanding of best pedagogical practices.

The NGSS practices are *science*-centric—they were distilled with a scientific lens and then tweaked so they would apply to *engineering*. Thus, they do not call out some of the most fundamental and distinct practices of engineering. In the second part of this chapter, I seek to provide additional support to educators interested in implementing engineering. I outline a set of engineering practices (which I call engineering "habits of mind") that more fully describe the kinds of work that engineers do and that should guide K–12 engineering activity.

ENGINEERING IN THE NGSS: OUT OF THE SHADOWS BUT NOT IN THE LIMELIGHT

The inclusion of engineering in national K–12 science standards is recent; the NGSS Frameworks (2012) and Standards (2013) are the first documents to use the term *engineering* explicitly. (For a thorough, readable history of STEM and engineering standards, see Sneider & Purzer, 2014.) Some states are choosing to adopt the Next Generation Science Standards—others are not, instead choosing to revise their state standards to align with the structures and content in the NGSS. No matter how states choose to adopt or adapt the NGSS, one thing is clear: Engineering is receiving unprecedented attention in today's K–12 districts and classrooms nationally.

Let's briefly consider the NGSS organizing structures. Three dimensions describe what and how students should learn:

- ***Practices***—what scientists and engineers do as they engage in their work. For example, both scientists and engineers *develop and use models*, and both engage in *argument from evidence*. The standards describe a developmental progression for each practice, beginning with kindergarten and increasing expectations through grade 12.
- ***Cross-Cutting Concepts***—concepts that apply across all domains of science and engineering, such as *patterns* or *systems*. Students are expected to learn to identify instances of these concepts in all areas of science and engineering, and understand them as themes that traverse the disciplines.
- ***Disciplinary Core Ideas***—a delimited set of core knowledge and ideas such as *weather and climate* and *growth and development of organisms*. These are the "facts" and "understandings" that science standards have more traditionally addressed.

It is important to recognize that science is the primary focus of the NGSS document; engineering is a minor part. Thus, the structure and features of the standards document were developed with science in mind. These Practices-Concepts-Ideas categories are distinct in science and they function well for this discipline. However, as explained in Chapter 2, engineering is a distinct discipline from science and the Practices-Concepts-Ideas structure doesn't map neatly onto it, which can create confusion for users of the NGSS. Educators trying to understand the complex, multidimensional framework and standards are likely to assume that the dimensions of science and engineering are organized consistently.

The NGSS's treatment of Disciplinary Core Ideas (DCIs) differs between science and engineering. Science is broken down into various disciplines (Earth, physical, and life science). Each discipline has disciplinary core ideas. The 38 science DCIs are concepts, principles, or theories such as "forces and motion" or "natural selection." In contrast, engineering is treated as one discipline with no specified fields. There is only one Disciplinary Core Idea unique to engineering: engineering design. This core idea describes important engineering processes that students should engage in:

- Defining and delimiting an engineering problem
- Developing possible solutions
- Optimizing the design solution

One reason the NGSS are confusing when viewed with an engineering lens is that they conflate engineering ideas, practices, and design processes (for a deeper explanation and critique, see Cunningham & Carlsen, 2014a, 2014b). The good news is that the document recognizes the importance of engaging students in engineering design as the critical feature of K–12 engineering education. The focus on process makes sense for K–12 engineering education. The takeaway message from the NGSS is that K–12 educators should focus on the processes and practices of engineering. In the next section, I elaborate on the engineering practices identified by the NGSS.

NGSS ENGINEERING PRACTICES

The NGSS emphasize building students' understanding of how to practice science and engineering and developing their abilities to competently engage in such practices. Scientists and engineers engage in a wide range of practices. The NGSS document specifies eight of these as foundational for science and engineering:

1. Asking questions and defining problems
2. Developing and using models
3. Planning and carrying out investigations
4. Analyzing and interpreting data
5. Using mathematics and computational thinking
6. Constructing explanations and designing solutions
7. Engaging in argument from evidence
8. Obtaining, evaluating, and communicating information

Science and engineering are both similar and symbiotic, with many overlapping practices. However, they are also distinct disciplines with different goals, as I described in Chapter 2. The NGSS discuss how practices vary by discipline, but substantively differentiate only two of them with respect to science and engineering:

- Asking questions (science) and defining problems (engineering)
- Constructing explanations (science) and designing solutions (engineering)

This distinction seems to suggest that the starting point, or goal, of these disciplines differs, as does their output, but the kind of work done in between is largely similar. This is not the case—important differences exist in the practices of engineering and science. In this section, I examine the eight NGSS practices through an engineering lens. I do this to clarify how science and engineering diverge and how are they are similar. Table 4.1 summarizes how each practice might manifest itself in a science and an engineering context, highlighting differences between the two disciplines.

If the goal is to engage students in authentic engineering experiences, we need to focus on engineering practices. Understanding how the NGSS practices manifest themselves in engineering contexts will help teachers become more confident as they support students during engineering activities. As I explore each practice in depth, I first compare science and engineering expressions of the practice and then I offer a classroom vignette to illustrate how elementary students engage in the practice as part of an engineering challenge.

Asking Questions and Defining Problems

The NGSS note the differences between the goals of science and engineering by delineating that scientists ask questions about the natural world with the goal of developing crucial explanations and theories about how the world operates whereas engineers aim to develop concrete solutions to real-world problems. For example, a scientist might ask how a cell becomes cancerous. An engineer might work to develop a chemotherapy pump that maintains the concentration of drugs at a constant level.

Solving engineering problems often begins with asking questions that further define the problem: What problem are we trying to solve? What do we know about the situation? What kinds of scientific knowledge might apply? What have others done? Engineers also define the criteria for successful solutions. What kinds of metrics does the design need to meet? Then engineers identify constraints and available resources that limit and support their designs. What materials, budget, or timeline considerations must be met?

Table 4.1. Differences in the NGSS Practices of Science and Engineering

Practice	Relative Emphasis in Science	Relative Emphasis in Engineering
Asking questions and defining problems	Goal is to theorize or further understand a concept	Goal is to create a useful, novel technology
Developing and using models	Scientists explain and predict	Engineers analyze and evaluate
Planning and carrying out investigations	Scientists hypothesize and test, often sequentially	Engineers evaluate, usually iteratively
Analyzing and interpreting data	Scientists attend to measurable aspects of the found, natural world	Engineers attend to diverse criteria: scientific (e.g., material properties) and other (e.g., cost, risk of failure
Using mathematics and computational thinking	Scientists test conceptual models with real data	Engineers design concrete things, using both real and simulated data
Constructing explanations and designing solutions	Objective is a single "best explanation"	Objective is a preferred design, selected from among alternatives, with explicit consideration of tradeoffs
Engaging in argument from evidence	Goal is to persuade scientific peers	Goal is to satisfy a client
Obtaining, evaluating, and communicating information	Free exchange of information is an important norm	Products are often legally proprietary, and information guarded

(based on Cunningham & Carlsen, 2014a, 2014b.)

Elementary Vignette

Ms. Brown has just challenged her 3rd-graders to design a lighting system using mirrors and a flashlight. This lighting system needs to illuminate hieroglyphs on the walls of a model ancient Egyptian tomb constructed from a copy-paper box. Ms. Brown models the practice of asking questions and defining problems by posing three types of questions. The first question type prompts students to think about the relevant background knowledge they might bring to the problem. Ms. Brown asks: "What do we know about light?" Students recall what they learned in their science unit about light—it can be reflected, blocked, and absorbed; it fades things; it moves in a straight line until it hits an object—and Ms. Brown records their responses on a chart. Then she prompts deeper thinking with two specific questions: "How does light interact with different kinds of materials?" and "How does light move?" These questions launch a series of student investigations of how various materials interact with light. Students reflect on their experiences and observations as a class to generate additional science information they can use.

A second set of questions helps students articulate the parameters of their problem. Ms. Brown asks: "What does our lighting system need to do? What requirements do we need to meet?" She documents the responses as "Criteria." These include lighting up as many details of the hieroglyphs as possible and shining as intense a light on the hieroglyphs as possible. Then students consider "Constraints": "What constraints are there on your lighting system design? What can we not change?" They identify the hieroglyphs, the location and size of the windows, where the light source goes, and how many light sources there are (one) as constraints they will accommodate.

Students generate the third set of questions. Ms. Brown invites students to share their questions about the project: "Now you get to ask me questions. We know what we are supposed to do. But what do you want to know about this project?" The students ask a range of questions that include the following:

- What kinds of materials are we using and how many?
- Do we get different light sources?
- What properties do the materials have?

Ms. Brown uses these questions to further define the problem and its criteria and constraints. She records them all on chart paper and then she answers each one, as well as any additional questions that arise. Students continue to ask questions throughout their engineering challenge, but these first questions serve to orient the students to the task at hand. (See video of this vignette at eie.org/book/4a.)

Developing and Using Models

Scientists and engineers use models in varied ways. Most science begins with conceptual models that can be refined through investigation. Scientists develop and use diagrammatic, mathematical, conceptual, or physical models to help them understand how the world works—to explain and predict. In science lessons, students often begin with a model and then engage in activities to better understand or refine this abstract concept. For example, students might draw a model demonstrating what happens when liquid in a flask sits in the sun to represent their understanding of how abstract concepts like evaporation and condensation work. After conducting investigations, they might update their drawings to reflect their new understanding.

Models are products or representations of engineers' investigations. Models help engineers design effective solutions to problems. Engineering models facilitate the analysis and evaluation of designs. As students manipulate models, they come to better understand design parameters, how materials function, and the interactions among various parts of the technology. They can also compare various designs and identify how they are failing. Constructing models allows engineers and students to bring a simplified version of a real-world technology into their lab or classroom. Professional engineers might construct physical models of a wide array of airplane propellers, varying a parameter such as the angle of the blade slightly for each, and then test them in a wind tunnel to see which works most efficiently. Similarly, in classrooms, students create physical models as they solve a challenge. In engineering, models are often the product of the investigation. By observing and tweaking these to better meet the goal, students can develop their understandings, present their ideas, and communicate their findings.

Elementary Vignette

Second-graders need to work as mechanical engineers to design blades for a windmill. Their teacher, Ms. Slater, shows them the small model windmill base design onto which they attach their blades (see Figure 4.1). Ms. Slater demonstrates how the windmill should function. Then she challenges her pupils to design the blades, reminding them to consider number, orientation, size, and materials. The students use the model to test various blade configurations and compositions and collect data about how well each design works (see the next section). They assess whether their blades spin and, if they do, how well the blades work by testing how much weight (how many pennies) the blades can lift. Using a simple desktop model allows students to test their solutions, explore concretely how relevant variables affect windmill performance, and determine which designs best meet the challenge's criteria. (See a video of this example at eie.org/book/4b.)

Planning and Carrying Out Investigations

Both engineers and scientists conduct investigations, but they execute these investigations for different reasons. Scientists plan and carry out systematic investigations to answer questions about natural phenomena. They aim to evaluate or extend knowledge about the natural world by producing an explanation that advances our understanding of the world. For instance, scientists who study hydrothermal vents in the ocean floor design investigations that allow them to identify the chemical composition of the fluids that spew from these underwater volcanos. They might also explore how microbes and organisms evolved to survive in these unique environments.

Engineers, however, conduct investigations to gather data that can help them specify design criteria and to evaluate how well their design meets specifications. They use this information to then develop a technology. For example, engineers might design a device to collect microbes. They would conduct investigations to identify criteria and constraints that will help them design chambers that can withstand the extremely high pressures that exist on the ocean floor. They might also figure out how to heat the water in the devices so the microbes can survive.

Figure 4.1. Model Windmill

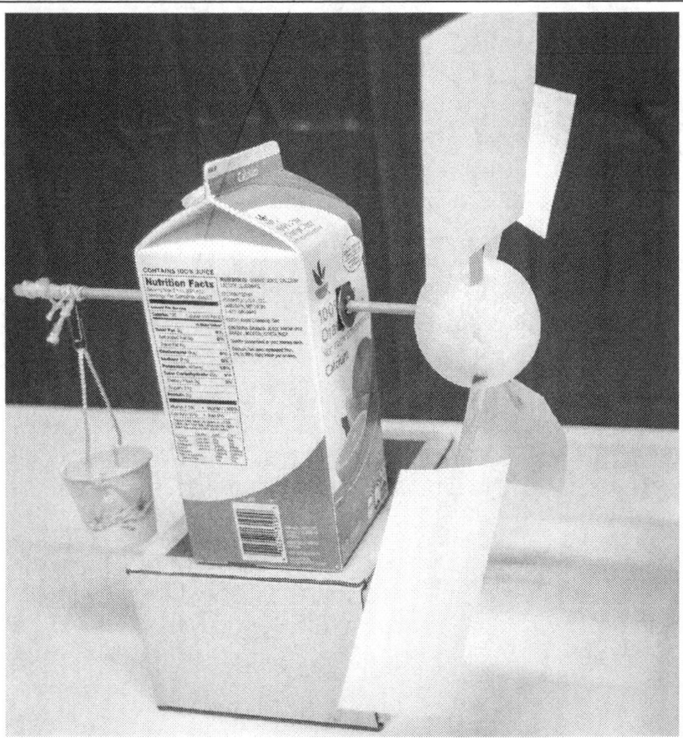

Both scientists and engineers identify what data should be recorded and what variables need to be considered, but the outcomes of the investigations differ. Engineering investigations usually have multiple acceptable solutions, while science aims to provide a stable description of the natural world. During engineering activities, students investigate many alternatives to find solutions that address the design problem criteria and balance tradeoffs in different ways.

Elementary Vignette

After Ms. Slater's 2nd-graders design their first set of windmill blades, they need to investigate if and how these windmill blades work. Jacob and Samuel have planned a blade design that uses three small paper cups. They have constructed it and tweaked it until they are satisfied, so they approach the testing station. Ms. Slater works with the team to investigate the capabilities of their blade design by assessing how much weight their windmill can lift. (View the scene below at eie.org/book/4c.)

[Teacher turns on fan. The windmill blades begin to spin. She turns fan off.]

Ms. Slater: So, we know it works. Let's see if it will also do some work. There is nothing in the cup this time.

Samuel: There's nothing in the cup, but see if . . . let's see if we put a penny.

Ms. Slater: Let's put five. Let's put five pennies. [Puts five pennies in the cup] Now let's see if it works.

Jacob: Now let's see if it works. [Turns on fan. Initially nothing moves. But then the blades slowly start to turn and the cup rises.]

Ms. Slater: See, it has to work harder, doesn't it?

Jacob: It goes up, it's going, it's going, it's going.

Ms. Slater: There it goes. Good job. Okay, let's try it again. See if we can put 10 pennies in. [Turns fan off. Puts an additional five pennies in the cup.]

Jacob: Ohhhh.

Samuel: That's not going to work.

Ms. Slater: Now we have 10 pennies in the cup. Do you think it [the wind] will spin it [the blades]? Do you think it will go faster or slower?

Jacob: Slower.

Samuel: Faster.

Ms. Slater: You think this wind is going to make this cup go up faster or slower?

Jacob: Slower.

Samuel: Slower.
Ms. Slater: Why? Why, Jacob?
Jacob: Because there's too much. They're way too much pennies.
Ms. Slater: And the weight is too heavy, right? And there's a lot of work to do, doesn't it? Let's try. [Turns fan on] Hmmm. It's struggling. It's struggling.
Jacob: It's struggling. It's struggling. It's struggling.
Ms. Slater: It's doing it. There it goes.
Jacob: It's doing it.
Ms. Slater: Is it going as fast as it was before?
Jacob: [Shakes head "no"]
Ms. Slater: How come, Stephen? How come it's not moving as fast?
Stephen: Because there's so many pennies in there.

Through rapid cycles of planning and investigation (testing) the 2nd-graders develop their understanding of how windmill blades can effectively capture wind energy.

Analyzing and Interpreting Data

All scientists and engineers analyze and interpret data—it's central to both professions. Scientists rely on data to generate evidence for new or existing scientific theories, whereas the data that engineers collect tend to focus on optimization—the process of ensuring that a design best meets a particular purpose within a set of constraints. In general, scientists collect data with the goal of producing generalizable claims. For example, scientists researching the Zika virus use data from laboratory science and epidemiological studies to learn more about how the virus has evolved and how it causes disease.

Engineers analyze and interpret data to evaluate the flaws and strengths of specific solutions and to better understand how the solutions can be improved. As they evaluate solutions, engineers consider a wide array of data that address the important goals and constraints—material properties, cost, environmental impacts, safety, and societal factors, among others. For instance, once a Zika vaccine has been created, industrial and manufacturing engineers will use data to design optimized processes and systems that maximize the vaccine production while adhering to legal, ethical, cost, transportation, packaging, and safety requirements.

Students engaged in engineering challenges are often highly motivated to understand the data they collect so they can improve their designs. Through analysis, students learn more about how the technology functions. Because they are honing their data collection skills, interpretation and analysis can also present an important opportunity to reflect upon their testing procedures and to discuss method and outliers.

Elementary Vignette

Mr. Lebel's 2nd-grade class has engineered three different kinds of bridges from index cards. They have tested them to determine which held the most and least weight by adding washers until the bridges collapsed. Mr. Lebel leads a class discussion to interpret the data. He requests that each group report which bridge type supported the least weight and how many washers it held. Three of the groups identify the beam bridge as the weakest. But one group's data differ. Their arch bridge was the weakest. Mr. Lebel invites the groups to discuss their findings. The group with the outlier data tries to understand their results. One student, Ruby, says, "That's kind of confusing to me. Because wouldn't it [the findings] be the same thing if . . . everyone does . . . [it the same]?" Another classmate, from a different group, interrupts her, suggesting, "Maybe they [the outlier group] threw [the washers] in and they got the least." This opens a conversation in Ruby's "outlier" group about their testing procedure, which leads the members to determine that they used two different methods for getting the weights into the cup—gently placing and throwing. They identify this as the source of their different results.

The class reconvenes and the members of the outlier group share their explanation with the larger class. They admit that they were initially throwing the weights in:

Ruby: Well, [we were throwing them in] at first. But then you [Mr. Lebel] told us. Then Victoria told us to drop them in light[ly], so we had to do it all over again.

Mr. Lebel: So, you redid your experiment again, so . . . this ended up being when you placed . . .

Ruby: Well, we didn't redo that [the arch bridge, the outlier] one. We did all the others.

Mr. Lebel: Wow . . . Wow. Wow. Wow. That is a huge realization. They just admitted, which is fine—thank you for admitting, I love honesty—that they redid the other experiments using light weight but they forgot to go back to the arch bridge and gently place it. Is their data off? Yes or no? Raise your hand "yes." Raise your hand "no." Did they do something different for one bridge than the rest of the bridges?

Students: Yes.

Mr. Lebel: So, can we use their data?

Students: No.

Mr. Lebel: No, 'cause it's not treated equally. That's a hard lesson to learn, isn't it?

As these young engineers interpret their data by comparing them with the larger set collected by their peers, they are highly motivated to figure out why their data are different. Their analysis provides them, and their classmates, with a valuable example of how to puzzle through outliers and the importance of consistent testing procedures. (View the video at eie.org/book/4d.)

Using Mathematics and Computational Thinking

Science and engineering both rely heavily on mathematics and computational thinking. Engineers use mathematics when testing the viability of possible materials and solutions and use their calculations to make informed design choices. Scientists use mathematics and computer simulations to analyze and describe their data, often calculating rates of change or relationships among variables. More complicated science experiments involving many interacting parts and randomness require computer simulations to see what a certain theory actually predicts. Engineers also use mathematics and computer simulations for these data analysis purposes, and to estimate the best designs rather than having to test all the possible designs. In elementary classrooms, students can use developmentally appropriate processes for measurement, data analysis, and calculation when they design and assess their engineering solutions.

Elementary Vignette

Fifth-graders working as geotechnical engineers are going to recommend a location for a gondola-type bridge called a TarPul (see Figure 4.2). To do so, they explore how soil type (rocky and organic) and soil compaction affect the strength of the foundation (how much weight it can support). Each group tests its soil sample three times and collects data. Then they must analyze the results of three trials. As they do so, their teacher, Ms. Hall, guides them through the process, reminding them, for example, how to calculate an average: "9 add 13 add 7 equals." Then she tells them to "divide it by 3" and asks her students "How come by 3?" The students, who are clearly already familiar with calculating averages, call out, "Cuz there's three trials." Ms. Hall instructs each group to perform the calculation on their own group's data. They first do the math individually. Then she instructs, "When you are all done, you need to compare it with your team to see if your [results] all match. What are you going to use to figure this out?" The students respond, "Calculator." Her students take out their smartphones, call up the calculator, and get to work. Once they have generated their numbers, Ms. Hall asks her pupils to think about what this calculation means for their design decisions: "Look at your data. What does this tell us about what you want to do with the soil? Julia, share with us what you think." Julia says that

Figure 4.2. TarPul Bridge

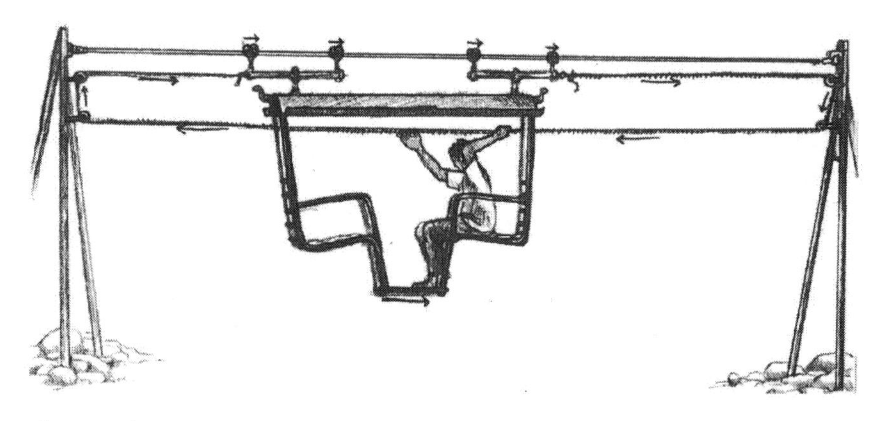

her numbers suggest that they should compact the soil more because the firmer it was, the more weight (people) the bridge design could hold. (View this video at eie.org/book/4e.)

Constructing Explanations and Designing Solutions

Science focuses on developing explanations of the natural world. Scientists strives to set forth a "best" explanation for an observed phenomenon. For instance, scientists can explain why the Aurora Borealis (Northern Lights) occurs. The outcome of engineers' work is a design that solves a problem. Chemical engineers, for example, may design a laundry detergent that is more environmentally friendly. As they develop these solutions, engineers will also need to offer explanations for their design decisions. Although engineers may examine data from a range of related problems to generalize, many engineers focus on suggesting solutions specific to a certain setting or situation. In classroom engineering, students embrace the opportunity to generate original design solutions and work diligently to improve their function.

Elementary Vignette

Ms. Murphy-Garcia's 4th-graders are putting their knowledge of magnets to work as they design a magnetic levitation, or maglev, transportation system. Maglev trains are popular in Japan, China, and Korea. These trains use magnets to lift a train above the tracks, which minimizes friction and allows extremely high speeds. For students' designs to function, the magnets on a small "train car" need to repel the magnets on the "track" so it "floats" (see Figure 4.3). One group's first test of its system does not initially function as expected. Peter observes, "Wait, it looks like it's attracting." As the

Figure 4.3. Magnetic Levitation (maglev) System

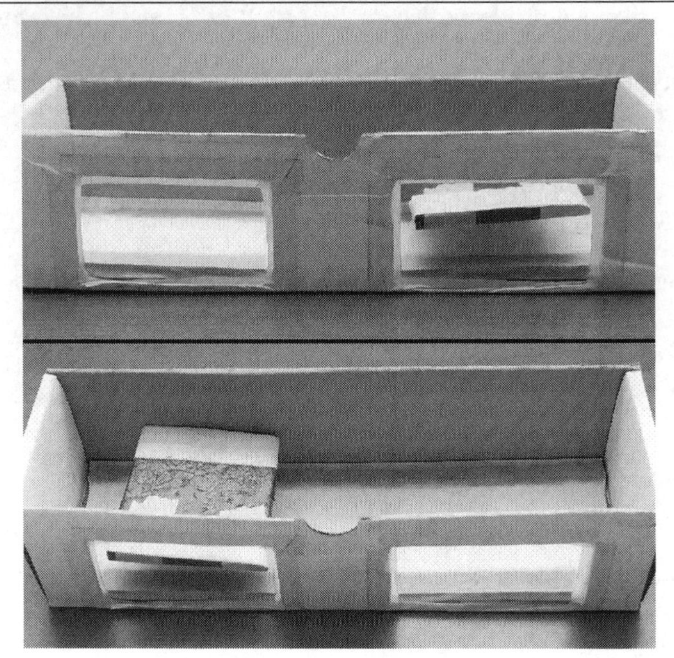

members of the group try to troubleshoot and figure out why it's performing in this manner, they offer explanations:

Peter: What I think is happening is this side here is like a different pole than this. So, it's trying to slide over. So it's on this one repelling but it's connecting on this one.
Ms. Murphy-Garcia: Okay, so . . .
Peter: It's trying to get inside there.
David: It's trying to go like that. It's trying to hop itself over.
Peter: If we made it a little bit wider and lost some of the magnets that run through the middle, it would work better.
David: When we improve.

The group redesigns the magnets on the train and tests again. This time, the solution works and the train floats the length of the track. The group exclaims, "We made it! We made it! We made it! Oh yeah. Oh yeah! High-fives to everyone." The pride and excitement is infectious! (View this video at eie.org/book/4f.)

Engaging in Argument from Evidence

Scientists engage in logical argument to propose new theories and to modify or strengthen existing ones. They make arguments to persuade their peers and the larger scientific community to accept their claims. For example, scientists continue to compile evidence that might answer the question of why dinosaurs became extinct.

Whereas scientists argue for specific theories and explanations, engineers engage in productive arguments to advocate for a specific design solution. They work in multidisciplinary teams, in which individuals with different expertise may propose varying solutions. They also aim to have a client, or their larger company, adopt, create, or use their technology. A team working to engineer improved smartphones must justify why its pet feature should be developed and included in the next version based on user experience, performance, cost, or efficiency. Because multiple possible solutions can exist for engineering problems, engineers must make tradeoffs between various criteria and constraints. Students, like engineers and clients in the real world, should present and argue for their design solutions using relevant evidence.

Elementary Vignette

Teams of 3rd-grade students from Ms. Keeling's class are also working as geotechnical engineers. They need to figure out where a TarPul bridge would be best situated along a river (see Figure 4.4). They know where their clients, the villagers in a story they read, want the bridge to be. They know also the geological composition of the Earth's layers (the bridge stanchions must be anchored) and the topography of the river (the riverbanks will erode faster along curves). Using this information, students must optimize these variables, choose a site, and convince others on their team that it is the best location. During their deliberations, students argue passionately about the advantages and disadvantages of the potential sites, which are labeled with letters. They draw on multiple forms of evidence as they consider the benefits and drawbacks of each potential site:

Ms. Keeling: Okay, it sounds like you have narrowed it down to [sites] E and F. Okay, I'm going to come back and check on you; you've got to narrow it down because a lot of groups are down here talking about the advantages and disadvantages, which you've already talked about. But we've got to get it down on paper like an engineer would.

Oliver: All right, I think we should do F.

Phillip: F.

[Noah nods]

Richard: What?!

Oliver: Three on one.

Richard: Let's do E.

Oliver: Why?

Richard: Because it's closer to the clinic.

Phillip: But it's closer to a curve.

Richard: F you have to walk to.
Phillip: It's closer to a curve.
Oliver: So.
Richard: Yeah.
Oliver: What are legs for?
Noah: F.
Richard: No. The Agee [grandmother in the storybook] cannot walk because she is sick.
Oliver: Yeah. So somebody will carry her.
Phillip: Or help her.
Noah: But she's well, after all.
Phillip and Oliver: She's well, she's well.
Phillip: Come on, E's closer and it's straight.
Oliver: And it's three on one.
Phillip: Three on one. Three on one.
Richard: That's it. We're doing rock, paper, scissors.

(Watch this interaction at eie.org/book/4g.)

Obtaining, Evaluating, and Communicating Information

Both scientists and engineers obtain, evaluate, and communicate information through text, graphs and other data, and diagrams and models. They also give and critique presentations and design prototypes. But in the real world, each

Figure 4.4. TarPul Design Challenge

group shares information in different ways. In science, knowledge and information is typically shared publicly with the whole research community. Generally, scientists publish academic papers or present at conferences to share their findings, which are then available to the scientific community. In professional engineering, solutions might be patented or proprietary. Often, such information can only be shared with a small group of senior members. For example, Apple, OxiClean, Ford, and Monsanto do not share the knowledge underlying their technologies with competitors.

However, in the classroom, engineering is often a more collaborative endeavor in which students share information through design briefs and presentations. Also, because design solutions are going to be adopted and used by other people (the clients), it is critical that the engineers explain how and under what conditions their design works in everyday terms.

Elementary Vignette

Ms. Hamilton's 3rd-graders are also working as geotechnical engineers. They have completed their planning, investigation, data collection, and analysis of the TarPul challenge. These engineers are now ready to recommend to the Nepalese village elders (their client) where they should situate their bridge. Each group communicates its findings through a report that summarizes the group's testing and recommendations. As a team, the students work together to write a persuasive speech that they will give. In her report and notebook (see Figure 4.5), Eleanor writes:

> We have selected site E. Why? Because we get how you feel about the location. The site we picked is only one sight [*sic*] away from where you want it [the TarPul] to be. The site you want is on a river bend and will erode more. The site we recommend is on a straight part of the river and will erode less. We can compact site E's soil ½ inch and it will make the poles holding the TarPul sturdy and strong. Where you want the TarPul is on rocky soil on one side and organic on the other—your TarPul will be leaning down on one side. We can compact the soil but it will not be as safe and sturdy as site E. Your TarPul will help you get across the river in the rainy season and you won't have to skip school so much.

The eight NGSS practices move the conversation about instruction beyond facts and knowledge to focus on how students construct understanding and make sense of their work by engaging in meaningful activity. But the practices suffer from a science-centric perspective because they were generated as fundamental practices of *science* and then modified slightly to accommodate engineering. However, when viewed with the lens of which practices are most

Figure 4.5. Persuasive Speech About TarPul Site

Lesson 4 Geotechnical Engineers' Persuasive Speech

सुमन सुमन सुमन सुमन सुमन सुमन

Write a persuasive speech to the village elders to explain to them why you think your selection is the best site for building the TarPul. In your speech, include the following points:

- Which site you have selected.
- Why you think this is the best location for building the TarPul.
- The amount you recommend compacting the soil around the TarPul foundation.
- Why you believe that this is the best amount of compaction for building the TarPul foundation.

¶[We have selected site E. ¶ Why? Because we get how you feel about the location. The site we picked is only one sight away from where you want it to be. [The site you want is on a river bend and will erode more] The site we recomend is on a straight part of the river and will erode less. We can compact site E's soil ½ inch and it will make the poles holding the TarPul will make sturdy and strong. ¶ The [Where you want the TarPul is on rocky soil on one side and organic on the other your TarPul will be leaning down on one side. We can compact the soil but it will not be as safe and sturdy as site E. ¶ Your TarPul will help you get across the river in the rainy season and you won't have to skip school as much.]

DM (4-10) 48

fundamental to *engineering*, it is clear that this is not a robust list. The second part of this chapter turns to explore features that define high-quality engineering in a classroom.

ENGINEERING HABITS OF MIND: PUTTING ENGINEERING IN THE SPOTLIGHT

When I began to think about introducing engineering to elementary students, I asked two questions: "What should engineering look like at the elementary level?" and "How will we know when high-quality engineering instruction and learning occurs?" I believed if we focused on those two questions, the answers would inform the lessons we produced and the professional development opportunities we offered. To answer those questions, I turned to multiple sources. I reviewed research studies that explored professional engineering in practice to build a more robust understanding of engineering and how it works. I asked engineers working in the industry, academia, and government what they thought defined their work and what they thought children and the public should understand about engineering. My team and I discussed at length what we wanted children to experience and understand about engineering—what were core ideas of engineering and what should students do? How might these manifest at different age levels? Only after we explored these important questions did we begin to develop our engineering activities. As we tested and modified these activities, we talked to teachers, observed their classrooms, and collected thousands of student engineering assessments and notebooks, which provided valuable insights into how children engineer.

Through this work, my team and I identified 16 engineering practices: our "Engineering Habits of Mind" (see Figure 4.6). We created this list to guide educators as they develop their understanding of what K–12 engineering could look like. We use this framework to anchor our lessons and activities, our professional development, our research, our assessments, our classroom observations, and our conversations. Though our practices overlap with those articulated by the NGSS, there are important differences. The NGSS practices stem from science. Because of this, the NGSS list lacks practices that are unique and important to engineering. For example, engineers envision and often create multiple ways to solve a problem, whereas scientists seek the most generalizable and encompassing explanation. Balancing criteria and constraints while exploring design possibilities is a foundational practice for engineering, but

Figure 4.6. Engineering Habits of Mind

EiE's Engineering Habits of Mind

Children who develop engineering habits of mind . . .	
Develop and use processes to solve problems	Investigate properties and uses of materials
Consider problems in context	Construct models and prototypes
Envision multiple solutions	Make evidence-based decisions
Innovate processes, methods, and designs	Persist and learn from failure
Make tradeoffs between criteria and constraints	Assess the implications of solutions
Use systems thinking	Work effectively in teams
Apply math knowledge to problem solving	Communicate effectively
Apply science knowledge to problem solving	See themselves as engineers

not critical for the creation of scientific explanations. Thus, the eight NGSS practices do not adequately describe the fundamental practices of engineering, especially those that differentiate engineering and science. The 16 practices we generated arose by asking what engineers do as they develop knowledge and solutions (Cunningham & Kelly, 2017a). Many of these practices manifest in science and other disciplines, but often in different ways. Because the NGSS use the term *practices*, to avoid confusion between the two lists we decided to use another descriptor and settled on "engineering habits of mind."

Our list is not comprehensive—there is a myriad of practices in which engineers and scientists continually engage as they work. Nor are the habits on the list unique to engineering—subsets of them are used in other academic disciplines, and some, such as teamwork and effective communication, are good life skills. Students don't engage with every engineering habit of mind during an engineering activity, but we find it useful to reflect upon which ones are present (and which are not). Students will need multiple, scaffolded experiences over time to develop facility with these habits. As they gain experience and maturity, their participation and engagement with these habits of mind will become deeper, more routine, and more complex.

In the remainder of this chapter, I dive into these habits by bringing them to life in a classroom setting. The 5th-graders in Kelly Thomas's class in Hollywood, Florida, need to solve a transportation engineering problem—the same maglev engineering challenge (see Figure 4.3) as Ms. Murphy-Garcia's students (EiE, 2011d). They're a great class to demonstrate our engineering habits of mind. Approximately 90% of the students in the school are Black or Hispanic. The public, Title I school faces significant challenges, as 92% of students are eligible for free and reduced-price lunch, 40% of the students read at or above grade level, 40% are meeting state standards in writing, 46% are at or above grade level in math, and only 26% are at or above grade level in science. But you wouldn't guess these metrics if you saw how these students engage and perform during this engineering unit. As students engineer, they develop habits of mind that will prepare them for success in STEM fields and other academic areas. With their teacher's help, Ms. Thomas's students engineer a maglev transportation system and actively develop engineering habits of mind. (Watch video of this class at eie.org/book/4h.)

Describing the Context and the Problem

All students should begin their engineering experiences by learning more about the problem they will address and its context. We have found that storybooks are extremely useful for this purpose. In Ms. Thomas's class, students in small groups read *Hikaru's Toy Troubles* (EiE, 2007), a story that follows a Japanese boy, Hikaru, and his family. Immediately, the students identify the problem: The family's toy store is losing business to a recently opened competitor's store—the family's store needs an innovative toy to regain business. As mentioned previously, *problem solving* is the defining characteristic of engineering (Sheppard, Colby, Macatangay, & Sullivan, 2006). Students, like professional engineers, should focus much of their work on designing solutions to a particular problem. Most engineering knowledge and solutions are developed in response to a set of parameters determined by the situation and the clients' needs. Engineers need to attend to these external parameters as they develop solutions—they *consider problems in context.*

Ms. Thomas describes how taking the time to situate the challenge in a larger context benefits her students:

> The storybook helps set the context for the design process . . . in that the kids are able to see how Hikaru works through the different stages of the design process. They are also introduced to what transportation engineering is. . . . At the same time, they are learning about a different culture and they find that really interesting and I think it makes it more real and [they] understand, wow, kids can solve design challenges, too, because Hikaru's basically their age.

Hikaru works to solve his family's dwindling sales by enlisting the help of a neighbor, who is also a transportation engineer. Together, Hikaru and his friends design an attention-grabbing maglev transportation system to move packages down a track and into customers' shopping bags. This innovation attracts new customers and helps save the family's business. Like Hikaru, students in Ms. Thomas's class need to engineer a similar system. Though the problem-solving methods and processes engineers use are diverse, K–12 teachers often articulate and use an engineering design process to focus student work and introduce students to organized approaches to problem solving. Ms. Thomas's students will use our engineering design process (as described in Chapter 3).

Considering Materials, Science, and Math

During the Ask step of the EDP, students *consider the properties* of magnets. The previous week, Ms. Thomas's students studied magnets and their properties in science class. A number of times during the engineering unit, Ms. Thomas asks students to recall what they know about magnets. Ms. Thomas explains to the class: "This is gonna be important because we need to fully understand all the properties of magnets before we can design our maglev train, which we will do tomorrow." Through a guided discussion, the students share that magnets have two poles, which can attract or repel, and that magnets are attracted to some metals.

Studying magnets the week before definitely prepared these students for their design challenge. Students, particularly younger ones, benefit from structured opportunities to reflect upon what they know about science and mathematics so they can make connections that inform their engineering designs. Just like practicing engineers, students draw upon, reinforce, and *apply science and mathematics concepts* as they analyze, design, and improve their creations (Bucciarelli, 1994; Jonassen, Stroebel, & Lee, 2006). Classroom engineering activities should communicate the interdependencies of these three disciplines. As students use science and mathematics concepts they have previously learned, or as they recognize the need to learn more science to boost the effectiveness of their designs, students' science and math knowledge is strengthened (see Chapter 7 for research results). Though it is important to draw upon science and mathematics concepts, engineers note that science is not the driver of engineering solutions—engineering is not just applied science. Rather, engineers construct their own knowledge, but in ways that draw from and apply their knowledge from these other disciplines.

Ms. Thomas tells her students that they will visit three stations where they will explore how magnets behave under certain conditions. The students investigate where poles are located on different types of magnets and what affects the strength of magnets' attractive and repulsive forces (see Figure 4.7).

After students have cycled through all the stations, Ms. Thomas asks, "Did anyone learn something today that they didn't already know about magnets?" One student responds, "That magnetic forces can go through things." Another student describes how one magnet could be aligned so it could levitate three other magnets.

After the discussion, Ms. Thomas prompts students to think about what they know about the science of magnets and consider the properties of these materials. Students complete a worksheet that instructs: "List two properties of magnets that will be important to your maglev system and explain why the properties are important." After students reflect on this question individually, Ms. Thomas asks students to share what they have written. She projects a student's response on the board and reads it aloud:

> Okay. Let me share with you what Natacha put. Natacha says, "Property one, the repelling of the poles." She said, "This property is important because if you don't put the same poles together the train would not levitate." So that's a property of magnets that's going to help her in designing her maglev system. Now let's look at property number two.
>
> She says, "You can use the bar magnets to figure out the poles. This is important because if you don't know which pole is which, how would

Figure 4.7. Investigating Properties of Magnets

you put the magnets on the right side to make them levitate?" So, she's thinking about those properties and she's thinking, "How can this help me when I'm designing my maglev system?" Thank you for sharing, Natacha.

Here, Ms. Thomas models how she expects her students to consider their designs. They need to draw upon what they know about the properties of magnets and use this knowledge to inform their engineering designs. It's a critical feature of engineering—professional engineers understand and consider materials and their properties. Performance, cost, aesthetics, and interactions among various parts are critical features that determine what materials engineers use to create the technology.

Identifying Criteria, Constraints, and Implications

Similar concerns about materials and properties also arise during conversations about the *criteria and constraints* that bound an engineering problem. Criteria are the requirements that the design must meet. Guidelines about the amount of weight the design must support, how much space it can occupy, the life span of the product, the preferences of clients, or the impact of a technology on the environment are the sorts of criteria that engineers consider.

Constraints are limits that restrict the design, such as money, time, space, performance abilities of materials, or cultural norms. As engineers design, they need to be aware of parameters and carefully balance relevant tradeoffs and tensions to optimize solutions. Engineers try to articulate as many criteria and constraints as possible before they begin a project because they can produce successful designs more efficiently when they have clear guidelines. Of course, more criteria and constraints inevitably pop up throughout the design process.

Similarly, student designs are enhanced when all class members have a clear understanding of the design challenge's criteria and constraints. Listing the criteria and constraints can help students recognize how the goal of the activity will inform their designs. Following the "what do we know about magnets" discussion, Ms. Thomas reviews the goal of the engineering challenge to students: They will design a magnetic levitation transportation system. The system needs to float above the track and move small weights (see Figure 4.3). Then she asks her pupils, "What do you need to know to do this?" The students generate a list that includes the following:

- How do we keep the train on the track?
- How many magnets do we need?
- What color will it be?
- How heavy are the weights?
- How long is the track?

Ms. Thomas points out that the students ask questions about the criteria and constraints of the problem just like engineers do. It's important to note as well that as professional engineers consider the problem in context and articulate criteria and constraints, they also should reflect on the impact and *implications of the solutions in the real world.* Once engineers generate a solution, they should assess not only whether their creation solved the problem but also whether their creation is technically, socially, and ethically responsible. For example, engineers who design maglev trains must make sure that their creations function as specified, but they should also consider societal and environmental impacts as well as durability, comfort, and aesthetics. Engineered products can advertently or inadvertently harm people (atomic bomb), animals (DDT), or the environment (river dams). Identifying possible implications of an engineered product should always be considered when articulating a project's criteria and constraints.

Engaging in Innovative Systems Thinking

As Ms. Thomas answers her students' questions, she makes it clear that the students will design part of a transportation system—they will design the vehicle (or "train") and also the track. She emphasizes that the system components must function together—the "train" must work with the "track." She models the system and describes other system limitations—for instance, the track and vehicle must both fit into a box, which she provides. By using the word *system* in her language and by drawing students' attention to the fact that what they design must work with other fixed elements, just as they would in the real world, she helps develop students' *systems thinking*. Systems thinking involves understanding part-whole relationships and how choices for parts of a system have consequences for the overall functioning of the whole system.

After another refresher about how magnets function, and with the goals and limits of the system clear in students' minds, Ms. Thomas encourages her students to think of original, *innovative designs*. Ms. Thomas (and the curriculum) asks them to brainstorm at least two possible solutions, saying, "We need to start brainstorming ideas. . . . Sample materials are at your table. After you've had a chance to explore these materials, you're going to come up with two different maglev system designs." Initially, students work individually and document their ideas on paper by drawing diagrams (see Figure 4.8). Ms. Thomas reminds students to view their designs as a system, saying, "Make sure in your diagram you include your vehicle and your track."

As she circles the classroom, Ms. Thomas commends those students who are labeling their magnet poles: "I see students who are labeling the poles on their diagrams and that will be helpful when you go to design it," thus gently prompting the class to recall and *apply their science knowledge*—the trains will only levitate if the magnets are aligned to repel.

Generating Multiple Solutions with Models

Ms. Thomas's students' designs are very creative. By encouraging students to generate at least two designs, Ms. Thomas helps her students understand that engineering involves *envisioning multiple solutions*. Brainstorming a plethora of solutions is an important part of engineering work. Practicing engineers know that there is almost never just one way to solve an engineering problem, and engineering designs are never "finished." There is always the possibility for a new idea, material, or approach. Encouraging students to think innovatively and asking them to engage in out-of-the-box thinking can free them from the expectation that there is only one "correct" solution to the problem.

Once students have at least two ideas on paper, Ms. Thomas instructs them to resume working with their partners: "Come together, decide what to do, make a plan, and list materials." For the rest of the challenge, the students work in teams to plan, design, and improve their solutions. Teammates work together, reviewing the ideas they imagined and synthesizing them into one plan. As they do so, they rely on *models and prototypes*, just as engineers would. Typically, models encompass features or traits of technologies, processes, or systems that allow one or more parts to be examined or tested under

Figure 4.8. Student Designs for Maglev Systems

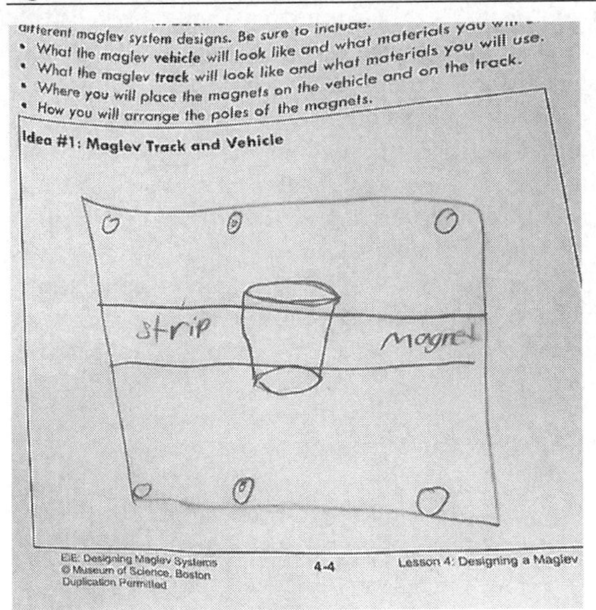

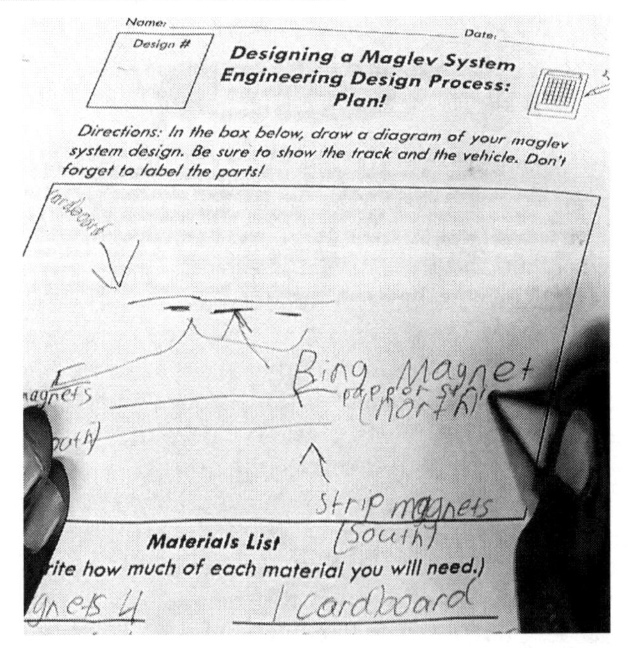

specified conditions—often under conditions when the object or process cannot be tested because of logistical, technical, or financial considerations. For example, Ms. Thomas's students (and engineers) cannot build a full-size maglev train to test.

The 5th-graders engaged in maglev engineering first create diagrammatic models to focus and document their thinking. They then construct a prototype of their system, arranging magnets on the vehicle. Models also allow engineers to make informed decisions and proceed ahead of predictive scientific theories. Engineers often rely on prototypes as a substitute for data, helping them make informed decisions (Madhavan, 2015; Vincenti, 1990). The students quickly find that engineering a maglev train is a challenging endeavor! Upon testing, most of their original designs do not perform as expected. However, Ms. Thomas's students are undeterred. As they test, they observe their vehicles' performance and notice where vehicles do not function as expected (see Figure 4.9). They keep trying and learning as they go. They make decisions about how to improve their system that are based in evidence, just as engineers do.

Learning from Failure Through Data-Driven Redesign

A key part of engineering is *persisting and learning from failure.* Engineering problems, tasks, analyses, and processes are rarely solved easily. Failure plays a prominent role and provides unique opportunities for learning and improving design (Madhavan, 2015). In considering the roles that failure plays in engineering practices, Petroski (2006) characterized success based on failure as follows: "Successful design, whether of solid or intangible things, rests on anticipating how failure can or might occur. Failure is thus a unifying principle in the design of things" (p. 5).

One way that engineers improve their designs is by *using data and evidence to make decisions.* Engineering is an empirical field that relies on interaction with the world, so data and evidence are crucial for effective engineering. The use of data permeates all aspects of engineering, from understanding the users' needs; to assessing the initial conditions of the problem or design challenge; to testing parameters of devices, components, or elements; to building effective prototypes; to testing models and designs in contexts; to presenting solutions to clients. Decisions need to be made at each juncture in the process. Often, engineering practices around such decisions involve empirical data, scientific knowledge, and uses of mathematical models to predict technical and commercial performance of different solutions (Trevelyan, 2010).

The students in Ms. Thomas's class engage in a rapid cycle of design–test–redesign. As they do so, their communication with one another and their teacher illustrates their thinking. Students explain their thinking to their partners, diagnose the problems with their current designs, and propose solutions while they work to improve their designs, as seen in the following quotes from various design teams:

Figure 4.9. Testing Maglev Systems with Weights

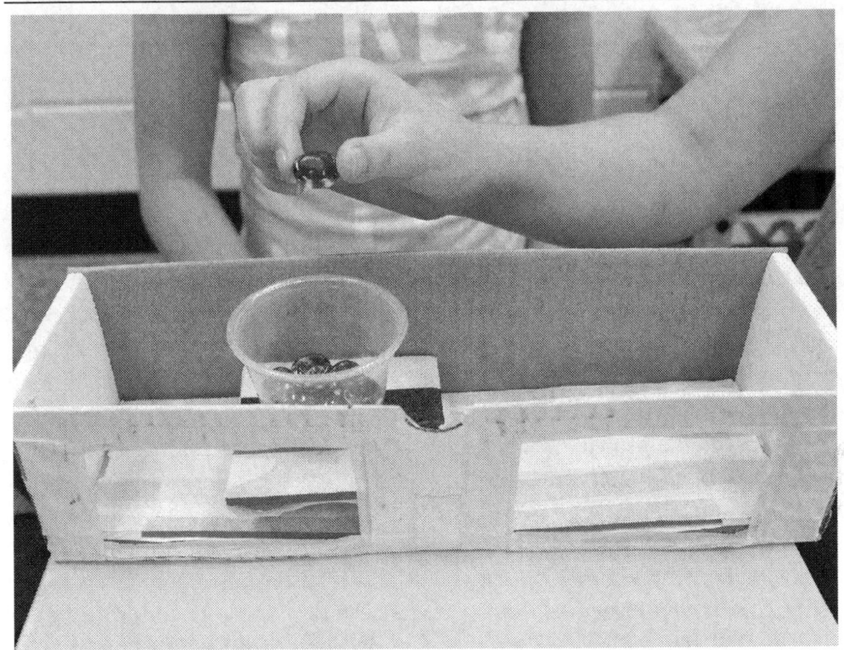

> [We are going to] Add more tracks. Or we could make this [the vehicle] more bigger [*sic*] so it doesn't have to move to the side. So it stops.
>
> We decided since we were putting two . . . of the magnet strings in the middle, we thought it was not going to work so . . . we're trying to put all the magnets in there so when it levitates it doesn't go to the sides.
>
> We need to make this bigger.
>
> We need to fix our train track.
>
> Definitely, [the vehicle] needs to be bigger.
>
> Try not to make any spaces on the side.
>
> We do need more of the strip magnets.
>
> [We are trying it] Without tape—maybe it [the train car] was rubbing against it [the wall] so we left it without the tape.

It is interesting to note a few things about these reflections. First, the students do not seem upset that their designs did not work the first time. The ability to face initial failure and still be motivated to try again is a valuable trait for students to develop. With engineering challenges, an Improve step is built into the design process. Students know right from the beginning that they get to try and try again—that engineering involves multiple iterations. When an initial design fails, that failure actually helps students learn more as they improve their subsequent designs. When failure is presented as a learning opportunity and an unavoidable step in engineering, failure is destigmatized. Students come to accept and even expect failure. Instead of worrying about being wrong or looking stupid, students come to focus on learning from their failed attempts. This is an important habit of mind, and a valuable life skill.

Second, these snippets of student discourse also demonstrate the diversity in the ideas for improvement that kids generate. Each has an original design. Their improvements will also need to be tailored to the behavior of their prototype. Some are focusing on materials, some are focusing on size, some on orientation, and some on construction. Each set of ideas is relevant to the unique design that the team has created.

Even when a group surprises itself by designing a successful technology, the group doesn't stop there. After one student realizes his idea works, he immediately considers ways to improve his design: "I guess . . . maybe we should put more than one wall." As the students' redesigns begin to succeed, the joy and pride they have in their accomplishments is infectious. Teams beckon Ms. Thomas to demonstrate their designs to her and they welcome the opportunity to share their results and designs with the larger class during the culminating whole-class shareout session.

Engaging in Teamwork and Communication

During this shareout, student give oral presentations of their designs. They describe the elements of their solution and demonstrate how their maglev system works. During the subsequent discussion, Ms. Thomas asks her students to reflect upon the engineering habits of mind that they used—specifically, *teamwork* and *communication*—and highlights these as important for professional engineers as well. Professional engineers recognize the influence that communication has on the efficiency and quality of teamwork. The processes of effective engineering design require that the designer be able to think and communicate across a variety of media and to a variety of audiences (Allie et al., 2009). Effective engineering communication requires making sure people understand designs and processes (Trevelyan, 2010). Although understanding technical diagrams and mathematical representations are part of knowledge production, other forms of communication such as "client interaction, collaboration, making oral presentations, and writing, as well as the ability to deal with ambiguity and complexity" (Jonassen et al., 2006, p. 146) are equally important.

Effective communication occurs in social contexts as engineers work together in teams. These teams often include other engineers, but also clients, technicians, artists, or even politicians. Engineers understand the importance of collaboration and the need to bring together expertise across types of knowledge. The design process requires negotiation and understanding of various interest groups, technical knowledge, and aesthetics.

Elementary teachers also need to help their young pupils learn to work together. Engineering activities that rely on teamwork can provide an opportunity for teachers to discuss explicitly the merits of teamwork for students and for engineers. Ms. Thomas herself does this with her class:

> ***Ms. Thomas:*** Do you think you would have been as successful if we hadn't shared as a group?
>
> ***Students:*** No.
>
> ***Ms. Thomas:*** Can you explain to me why? Malik?
>
> ***Malik:*** If we didn't get any suggestions then we'd, like, keep making ideas that like wouldn't work. Mistakes.
>
> ***Ms. Thomas:*** Okay. And not. And I wouldn't even call them mistakes because you don't know until you try it, right? You build on each other's ideas. Do you think you would have been as successful if you were working by yourself?
>
> ***Students:*** No.
>
> ***Ms. Thomas:*** Engineers never, Michael, they never work by themselves. They are always sharing with each other. And it's not copying. It's . . . it's learning from what other people have done and improving on their designs.

Developing an Engineering Identity

Ms. Thomas uses this discussion to connect what her students and engineers do. Engaging in engineering design challenges can help students recognize their potential as engineers and problem-solvers. Engineers, in classrooms or in professional settings, *develop identities* as practitioners with standards for quality and ethics. In their study of engineering, Anderson, Courter, McGlamery, Nathans-Kelly, and Nicometo (2010) found that "Engineering identity is a complex equation that factors in problem solving, teamwork, learning, and personal contributions. All of these combine to create a positive engineering identity" (p. 170). This identity is instilled in the community through the initiation of novices into the field. Many engineering education studies mention the importance of developing an identity as an engineer, and how this contributes to learning about the values of engineers, perseverance in the field, and learning to be a member of the community (Allie et al., 2009; Anderson et al., 2010).

From their first interactions with engineering, children begin to develop an engineering identity. Teachers play an important role in nurturing this growth. For example, throughout the maglev experience, Ms. Thomas refers to her students as "engineers" and comments often that they are "engineering" their designs. She draws out similarities between her students' work and that of real engineers. All of this invites her students to develop an engineering identity.

As my team and I design engineering resources and experiences for teachers and students, we think carefully about and articulate potential positive outcomes. One type of outcome focuses on what students should learn. Engaging in authentic engineering challenges affords students the opportunity to participate in the kinds of work engineers do. As students tackle new challenges, they hone problem-solving skills, learn to situate what they are doing within a larger context, and define the parameters and specifications under which they work. Students can be encouraged to draw upon previous knowledge as they think creatively. They may recognize the power of models and prototypes and the need to persist when an idea does not work as expected the first time. Interactions with their teammates, classmates, and teacher help them learn that creating knowledge and technologies is a highly social endeavor—they must work with others, support and justify their ideas, and communicate their thoughts and knowledge. Over time, engaging in a range of engineering activities that develop engineering practices will help students develop habits of mind that they can use throughout their schooling and their lives.

This chapter's deeper dive into the NGSS practices and engineering habits of mind illuminates key intersections between science and engineering and also highlights where their different goals and purposes cause their practices to diverge significantly. I offer the vignettes and video links mentioned above as brief glimpses into how these practices might manifest as students engineer. Of course, practices are complex, nuanced, and overlapping. Recognizing this, my team and I have created a library of additional video snippets culled from an array of engineering challenges in dozens of diverse classrooms across the country. We sort them by NGSS practice and by engineering habit of mind to show a range of ways that the engineering practices are developed through classroom activity. (To view them, please visit eie.org/snippets.)

As teachers develop more accurate, refined, and discipline-specific understanding of engineering practices and habits of mind, they can support their pupils as they engage in the engineering design process. Increased familiarity and experience with engineering enables teachers to guide classroom interactions, pose effective questions, and offer student feedback that encourages redesign and reflection. By engaging in engineering practices and habits of mind, students begin to understand how engineering solutions "come to be" and they can explore their potential as problem-solvers and engineers.

CHAPTER 5

Engineering Design Principles for Inclusivity

A commitment to promoting educational access and equity and closing achievement and opportunity gaps has grounded my work in education. Engineering should engage *all* students. I believe individuals, engineering fields, and society all benefit when everyone engages in engineering design, problem solving, innovation, and inquiry.

I particularly wanted to attract and engage underrepresented, underperforming, and underserved students, including girls, students from races and ethnicities underrepresented in STEM, students from low socioeconomic backgrounds, students who received special education services, and English learners. I decided one way to do this was to design curricular materials that reached these students. To guide my work, I began by identifying inclusive design principles that my projects would follow. I realized I had an incredible opportunity to introduce a new discipline to students without worrying about previous experiences or entrenched models. Instead of having students read about engineering in a textbook and then complete a worksheet, we could design activities that required students to do engineering in an engaging and hands-on way.

With equity as our focus, my team and I drew from educational literature and our experiences in classrooms to articulate a set of 14 design principles for engineering curricula and resources. These principles can be grouped in four larger categories as illustrated by Table 5.1.

Then we put these design principles to the test—working in concert with our habits of mind, these principles guide the development of the five engineering curricula we have created. Not surprisingly, the curricular design principles and the habits of mind overlap. In this chapter, we'll revisit some of the habits of mind we explored in the previous chapter, this time considering how those habits of mind engage populations underrepresented in engineering fields.

In this chapter, I examine *how* to design elementary engineering activities and curricular materials so they interest and engage the diverse students who attend our nation's schools using these inclusive principles. I describe them,

Table 5.1. Curricular Design Principles for Inclusivity

Category	Design Principles	Habit of Mind
Set learning in a real-world context	Use narratives to develop and motivate students' understanding of engineering's place in the world.	X
	Demonstrate how engineers help people, animals, the environment, or society.	
	Provide role models with a range of demographic characteristics.	
Present design challenges that are authentic to engineering practice	Ensure that design challenges are truly open-ended, with more than one "correct" answer.	X
	Value failure for what it teaches.	X
	Produce design challenges that can be evaluated with both qualitative and quantitative measures.	
	Cultivate collaboration and teamwork.	X
	Engage students in active, hands-on, inquiry-based engineering.	
Scaffold student work	Model and make explicit the practices of engineering.	X
	Assume no previous familiarity with materials, tasks, or terminology.	
	Produce activities and lessons that are flexible to the needs and abilities of different kinds of learners.	
Demonstrate that everyone can engineer	Cultivate learning environments in which all students' ideas and contributions have value.	
	Foster students' agency as engineers.	X
	Develop challenges that require low-cost, readily available materials.	

show how they can influence curricular design decisions, and provide teacher narratives that illustrate their potential impact on a diverse array of students. I offer these with the hope that they can also influence the work of educators who are interested in designing inclusive engineering lessons or curricula. Research studies and classroom implementation show that curricula that use these principles can improve engineering and science learning outcomes for *all* students (see Chapter 7).

CATEGORY 1: SET LEARNING IN A REAL-WORLD CONTEXT

Research on K–12 science education finds "many students who are academically competent in the school subject matter ultimately view school's knowledge and skills as irrelevant for their future career and/or everyday lives" (Carlone, Haun-Frank, & Webb, 2011, p. 460). To promote student interest in engineering and science as disciplines of study and as future career opportunities, educators should help students see how the topics they are learning are relevant in the real world. As mentioned in the previous chapters, students also need to be able to picture themselves filling such roles in the future. This is especially important for students from populations traditionally underrepresented in STEM.

Putting learning in a real-world context increases students' engagement, enthusiasm, and achievement. The three design principles that set a real-world context for all students, but particularly for groups underrepresented in engineering, are described below.

Principle 1: Use Narratives to Develop and Motivate Students' Understanding of Engineering's Place in the World. A compelling narrative engages students as they develop their understanding of an unfamiliar discipline like engineering. Narratives spark students' curiosity and imagination. We've found that providing a narrative context can especially boost the participation of girls and other underrepresented students who may be less predisposed to identify with or apply themselves to the physical sciences or more technical studies if they are unable to connect to the work on a personal level. When science educators, for example, teach science without grounding the work in a real-world context, often students wonder, "So what?"

Narrative stories and controversies that address science and engineering as they apply to people's lives have the potential to powerfully affect students' attitudes toward and understanding of science, engineering, and technology. Connecting school learning to events in the real world can motivate students by providing relevance for what they are learning, whether that is science or technological problems and problem solving (Buxton, 2010). When students better understand the significance of what they're learning, they're more motivated to engage and tackle the challenges in front of them.

Every engineering unit we create begins with a narrative that situates the engineering task students will do in a real-world context. We design our narratives to be age-appropriate and aligned with the setting in which they will be used. For example:

- In our pre-K curriculum, Wee Engineer, a puppet introduces children to the engineering challenge (Figure 5.1). The children learn more about the challenge they will tackle as the puppet walks

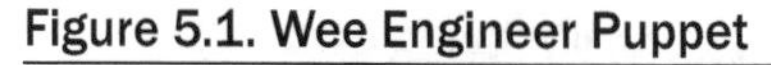

Figure 5.1. Wee Engineer Puppet

them through a problem it is trying to figure out and asks the children questions—for example, "I need to figure out how to make a loud noisemaker so that I can make a loud noise at my friend's surprise party. Can you help me?"

- Because reading is fundamental to elementary classrooms, in our classroom-based elementary curriculum, *Engineering is Elementary*, each unit begins with a short work of fiction written expressly for that unit—an illustrated storybook, set in a country around the world (Figure 5.2). This story introduces a problem that is similar to the one students will need to solve. In the storybook, a child protagonist solves the problem with support from an adult mentor who is an engineer. The global nature of the stories permits us to link to a diversity of students' backgrounds and experiences.
- In our elementary out-of-school curriculum, *Engineering Adventures*, a brother-sister team, Jacob and India, travels around the world (Figure 5.3). They kick off each lesson by sending emails, postcards, or audio notes. These notes share relevant information and encourage kids to help solve a challenge.
- In our middle school out-of-school curriculum, *Engineering Everywhere*, each unit begins with a "Special Report," a 10-minute video that highlights how professional engineers are working to solve

a problem closely related to the one kids will tackle (Figure 5.4). (You can watch these Special Reports at eie.org/special-reports.)

When we designed the stories for each EiE unit, we considered background science and engineering information that we wanted students to be exposed to (we would refer back to this during the lesson activities). We constructed a large table for the storybook characters. We wanted to make sure the characters came from all over the world, were equal numbers of boys and girls, were of all different races and ethnicities, and also included students with cognitive and physical disabilities. We made sure that the characters had a variety of hobbies and traits that students might identify with. They came from different family structures. We thought carefully about the storyline, the points of tension, and other attributes of high-quality children's literature. And though the child in the book does the same challenge kids in class will do, we were careful not to show a definite solution in the book—we did not want to influence students' designs.

Figure 5.2. EiE Storybook

Figure 5.3. Jacob and India

Students relate not only to these characters, but also to the settings. Often, we hear from teachers that students from other countries or cultural backgrounds became stars after the class reads a story set in their birth country—they are experts, teaching their classmates more about their native language or customs. Setting the stories in countries worldwide provides learners from different countries with the opportunity to relate to the content in a different and fulfilling way. For example, one 2nd-grade teacher from Fitchburg, Massachusetts, recounts below how the setting and narrative context of one of our stories engaged one of her English learner students in a very personal way:

> I just had to share that we began the engineering unit this week and my class is so engaged. I have 12 English learners out of 25 students total, one of whom is from China.
>
> The EiE storybook, *Yi Min's Great Wall*, which is set in China, really excited him. He was a socially awkward child, very reserved in class, he really keeps to himself, and the kids had a hard time relating to him. He was rarely excited about topics covered in class.
>
> When we started taking about the Great Wall of China, he knew all about it. He shared information about the Chinese language and about his family. He told us how to pronounce the Chinese word for "rabbit"

Figure 5.4. Special Report Video

[a word included in Chinese in the book], he was showing us other Chinese symbols, and the kids learned other words in Chinese from him.

For once, the other kids didn't see him as someone weird. They were asking him questions, and he really came alive. It really boosted his self-confidence, that he knew something that he could share and teach us! It was great to see him get so excited about something and be able to be looked to as an expert. It was amazing to watch that transformation. I wish I could have videotaped him poring over the storybook; it was something to see.

Principle 2: Demonstrate How Engineers Help People, Animals, the Environment, or Society. Research finds many students, but particularly girls and students from racial and ethnic groups underrepresented in STEM, are interested in "helping" careers (Miller, Blessing, & Schwartz, 2006). Many girls and young women want to see how school subjects are relevant to their lives and add value to people, animals, the environment, and society. At the grade-school level, most engineering activities still focus on the physical sciences—typical challenges include designing robots, model race cars, egg drops, or catapults—with no obvious connection to how these challenges aid people and society. Students of both genders hold stereotypical views about physical science being "for boys" and biological science being "for girls." Girls prefer and choose to participate more frequently in biological sciences than in physical sciences and engineering. As a result, many college-bound high school girls tend to choose people-oriented majors within the STEM fields. They are more likely to focus on disciplines such as biology, which they see as allowing them to better help animals or people, particularly through health-related professions.

We suggest that classroom activities be designed specifically to highlight the positive societal impacts of engineering. For example, electrical engineering is a field that has not traditionally attracted large numbers of women. Instead of presenting young students with a decontextualized challenge of creating an electrical circuit, we encourage educators to set design circuit challenges in a real-world context to better enable students to recognize the many ways in which electrical engineering helps society. One of our electrical engineering units challenges students to design a circuit as one component of a larger alarm system that alerts a caregiver when the water level in the drinking trough for baby lambs gets low. Students are able to see how their work (and by extension, electrical engineering) benefits the caregiver and the lambs.

Often, we hear from teachers that this "helping" context has the power to draw in students. For example, one 5th-grade teacher from Garden Grove, California, taught our *Shake Things Up: Engineering Earthquake Resistant Buildings* unit (EiE, 2016). Her students had been studying Earth science. Living in California, she recognized the relevance of learning more about earthquakes and chose to introduce engineering. The unit's design challenge is set shortly after the massive earthquake that hit Haiti in 2010. Around the same time, an earthquake of a similar magnitude had struck California, killing 25 people. The teacher describes how this particular unit motivated her students to solve more than one problem:

> When the kids researched Haiti and learned that 450,000 people perished—and many kids their age were orphaned—there was shocked silence in the room. Then my students started to wonder, "Why? Why such different outcomes in California and Haiti?"
>
> [To make their designs students] brought in everything you could think of, including mud and adobe—because that's what buildings are made from in Haiti! But on the day we did the "shake test," every single model failed miserably. Ordinarily, when my students fail a test, kids are crying, but this time, they were like, "Okay, Plan A didn't work. Let's go to Plan B." They weren't looking to me to solve things. They were looking to each other. They truly [became] engineers.

Without any prompting, many of her students went home and redesigned their buildings, working with their families to engineer a successful earthquake-resistant building. Her students became invested in not only engineering an effective earthquake-resistant building, but also in helping Haitian children who were affected by the 2010 earthquake. To help nurture her students' curiosity, the teacher reached out to a nonprofit called Hope for Haiti. It was through this nonprofit that her class was able to learn of a young Haitian boy who sold bracelets on the streets of Haiti to support his family. When the class's initial idea to help support this young man by buying his bracelets was not possible, her students once again became problem-solving engineers and decided to make their own bracelets to sell for him:

> When they told me they planned to charge 10 dollars a bracelet, I thought no one would buy them. But these kids had passion! They went around to all the other classes to promote the project and made almost $2,000 in 2 weeks. Global awareness was born. The lesson motivated and inspired my students to continue to problem-solve and think about others outside of their own world. This engineering activity pushed them out of being so egocentric and allowed them to care about other people and other places. Engineering challenges in international settings are the foundation of EiE units, and that framework allowed my students to make global connections.

Principle 3: Provide Role Models with a Range of Demographic Characteristics. Girls and students from racial and ethnic groups underrepresented in STEM, as well as many other students, often perceive science as difficult, uninteresting, and resulting in an unappealing lifestyle. In part, this is because of the stereotypes and mixed messages about engineers and scientists conveyed by the media—for example:

- Women/African Americans/Latinos/Hispanics don't like, or aren't good at, engineering or science or math;
- Scientists and engineers are nerdy, antisocial, and not attractive;
- Women don't like dirty or technical pursuits;
- Scientists (and engineers) are old, frizzy-haired, White men in lab coats;
- People with physical disabilities lack what it takes to be a scientist or engineer;
- Women need to think about having a family—and science and engineering are really difficult and all-life-consuming; or
- Men are more logical and rational. Women are emotional.

Presenting a variety of positive role models in media and curricula can combat these messages.

Our curricular design choice to feature storybook characters from diverse backgrounds does have a tremendous impact on students. Teachers who have taught our units tell us that it's common to hear a child exclaim, "[Storybook character name] looks like me!" or that a certain protagonist is his or her favorite. Students benefit when they can see engineering role models with similar backgrounds and attributes. A 2nd-grade teacher in Palm Springs, California, described her students' response in this way:

> I recently gave my class a survey about what subject they like best. I've been teaching for 17 years and the number-one answer is always "recess." This time one child chose recess, two chose reading, five chose math, and 13 chose science! I believe that is the first time I've seen science even in the top five, let alone as number one, and I am absolutely sure it is because of their engineering experiences through EiE units. My second-language learners are eating up the hands-on learning approach. Javier [a Hispanic storybook character who lives in Texas with a blended family] in the *To Get to the Other Side: Designing Bridges* unit speaks Spanish, looks like them, and has a blended family. Many of my students live with one parent, a grandparent, have stepparents and siblings, or come from the foster-care system: Javier was familiar to them. They learned the vocabulary and science content, designed and tested bridges, and became less afraid to fail. I have done other units and I love how the real-world, multicultural connections bring what I am teaching to life for my students. So fun to be a teacher!

In addition to depicting storybook protagonists of different sexes, races, and ethnicities, we also feature children with a range of disabilities: Despina uses a wheelchair, Kwame is blind, Michelle has Down syndrome, and Paolo has a congenital limb defect. Teachers indicate that these characters often prompt class discussions about the inclusion and needs of such students. Teachers of students with special needs or disabilities describe how their students appreciate role models that are like them. One teacher from New York City described her experience with our unit that engineers a playdough process. The storybook is set in Canada and features Michelle, a girl who helps her hockey team's fundraiser by selling sculptures made of playdough:

> I have never seen a children's book highlighting a character with Down syndrome, so *Michelle's MVP Award* caught my eye. This year I taught a grade 3–4–5 blended special education classroom and 80% of my students had Down syndrome. I used the book and unit to introduce them to engineering. The students really loved that the character was like them; we discussed this similarity and two girls chanted, "Like me, like me, like me." I also appreciated that the storyline modeled kids with and without disabilities working together.
>
> I teach at an inner-city school in the Bronx where most kids come from the Caribbean, so they had no idea what ice hockey [one important story element] was. To give them some background, I wore a hockey jersey and showed them video clips of the game. This left them asking if we could go watch a game. Instead, after we read the book, they designed playdough. The tactile nature of this activity worked particularly well for my students, many of whom are working on strengthening their

fine motor skills. Like Michelle in the story, we decided to sell our play-dough to other classes. My students made up little bags and the other 5th-grade teacher on the hall let my kids visit their classroom. Many of my students have speech issues, which include difficulty speaking in full sentences. So, first they practiced describing their dough. They then displayed their dough for the gen-ed [general education] students (their customers). Each of my students got a chance to explain why their playdough was the best: the softest, the prettiest color, the most improved. The other kids were cheering them on. They sold the bags for a dollar each. As for the money from our sales . . . it funded a class pizza party!

CATEGORY 2: PRESENT DESIGN CHALLENGES THAT ARE AUTHENTIC TO ENGINEERING PRACTICE

Research indicates that students' learning is more profound when they engage in realistic disciplinary practices and put key concepts into productive use (Duschl, 2008; Kelly, 2011). Students learn concepts and skills through experience as they work and learn in rich contexts that mirror disciplinary problems and practices. Recent research has shown that inquiry-based and project-based learning approaches in science lead to comparable or improved outcomes for students compared to more traditional lessons (Kelly, 2014). The next five design principles address students' need for agency and ownership and engineering work that is "real."

Principle 4: Ensure That Design Challenges Are Truly Open-Ended, With More Than One "Correct" Answer. Open-ended challenges with more than one "correct" answer can shake up established classroom dynamics. Too often, traditional science lessons present tasks where the students' goal is to solve a problem and achieve the "correct" answer. By mid-elementary school, students readily categorize themselves as "good at school/science" or "bad at school/science" based on how quickly they produce the right response. Open-ended challenges are particularly refreshing for students who underperform in a "find the correct answer" system. Often, these students are re-engaged and excited by the opportunity to think creatively and escape the tyranny of one correct answer. For teachers, open-ended challenges invite the use of new pedagogical strategies and ways of thinking about learning that emphasize the processes of learning and divergent thinking.

As described in the previous chapter, in the real world, engineering problems rarely have one unique solution. To model this aspect of engineering for students, design challenges should be constructed intentionally to allow multiple solutions. Each design challenge should articulate criteria and constraints, and students should evaluate their solutions in relation to how well they meet these. Students should be encouraged to take risks and experiment with out-of-the-box ideas. They should also have time to review their classmates' designs and reflect upon how they are similar to and different from their own.

The opportunity to design something unique makes engineering highly interesting and engaging for students. Both students and teachers appreciate that multiple solutions are possible—in fact, teachers say it's a highlight of teaching engineering. Ms. Mock, the 1st-grade teacher from Minnesota whose students engineered hand pollinators, called out the multiplicity of students' solutions as one of her favorite parts of the experience (watch the video at eie.org/book/5a):

> I think the variety of solutions that the kids came up with was the most exciting for me. None of the hand pollinators looked the same. Not even similar. They used, obviously, the same materials. They are very creative. They are very different. I think the most exciting part was to watch them try it. And then to see the wheels turning and to talk amongst themselves about how to improve, it was priceless.

Initially, students who have "mastered" the correct answer approach may not embrace these types of challenges. They may feel overwhelmed or uncertain, but we've found that this will lessen over time. During open-ended challenges, all students more readily make connections with their interests and prior experiences, and they have more reason to listen to and respect the views of others when all answers and opinions are valued.

Principle 5: Value Failure for What It Teaches. Embracing failure can be a new experience for students. But failure is a necessary and inherent part of engineering (and life). When a design fails, the design isn't "wrong." The engineers don't need to scrap the entire design and start over—this failure is an opportunity for engineers to improve their design. Learning that failure is valuable can be liberating for students, especially for students who have been low achievers on traditional academic measures.

Engineering activities should invite students to think about engineering as an iterative event, present failure as an unavoidable step, and require solutions to be redesigned at least once. Initial designs may "fail" to meet the baseline criteria, but these failures help students learn and improve subsequent designs, just like working engineers do. Often, teachers report that their students readily embraced the opportunity to keep honing their solutions, asking if they could continue to work on their designs before and after school, during recess, and over holidays: "Several students kept bringing in playdough that they made at home. They wanted to improve until they perfected it. When they brought it in, many wanted to share the improvements they made and why they felt it was successful" (grade 4 teacher).

When an initial design fails, that failure actually *helps* students learn more as they improve their subsequent designs. Through well-designed activities and teacher facilitation, failure can be destigmatized and students can come to accept or expect failure as part of a problem-solving process. More important, such experiences help students learn to persist through failure. A 5th-grade teacher in Bremerton, Washington, captured the importance of persisting for her students who "come from a large free to reduced lunch area. I think we're about 80%. And most of my kids struggle with giving up." She explained why they enjoy engineering and why this is so critical for them (watch the video at eie.org/book/5b):

> I think they like the interaction, and the hands-on, and being able to try again. A lot of these kids, they struggle with failure, and are often told they can't try again. And being able to try again and hearing they really enjoy that fact that they get to, "Oh, I get to start it all over. That's okay?" So, I think that's one of the really big things that pulls it. . . . I think that when kids are working with EiE, and they're working on problem solving, and they're developing their persistence, that the idea of not giving up and trying again and again and doing it again, so you improve it and that you keep going. So, it [the engineering curriculum] really develops that habit to not give up and to keep going and to go on and do things like go to college, and graduate from high school—you know, things that a lot of kids in this school often don't think of.

Working engineers often test their designs to the limit in the hopes of understanding under what conditions designs will fail. Embracing failure also can be highly engaging for students, who identify the opportunity to fail and make things better as one of the highlights of their engineering experience:

> "I like the plant project because it was fun to mess up and try again."
>
> "Now I know how engineers feel when things they design don't work the first time, but I still want to be one."
>
> "What I especially liked was when we got to improve our ideas to make our system better."

Instead of worrying about being wrong, with well-designed engineering activities, students can focus on learning from their failed attempts. A *student* doesn't fail; a particular *design* fails—and when it does, the failure sparks a set of new, improved ideas. The importance of scaffolding students through failure is recognized by a 4th-grade teacher in Kittery, Maine, as she reflected on the benefits of engineering for her students (watch the video at eie.org/book/5c):

> I'm the gifted-and-talented specialist in our school district, and failure is not something that a lot of my students experience on a regular basis. They often experience great frustration actually, because it's not something that they get used to. I try to emphasize to them all the time that when they're failing or struggling, literally their brains are growing. That's when growth happens is when there's failure.
>
> The fact that we went into this [materials engineering lesson] saying, "We're going to destroy these [walls] and they're all going to fail," I think on some level was liberating, especially for the kids who aren't used to dealing with that, and seeing that as that is your opportunity to improve. That failure is your chance to see how you can make it better. I don't think there was any part of the curriculum that was more important, actually, than that particular lesson.
>
> Overall, I think they really enjoyed seeing the failure and then being able to think about how they would do it differently.

Principle 6: Produce Design Challenges That Can Be Evaluated with Both Qualitative and Quantitative Measures. Just because a design challenge has multiple solutions does not mean every solution addresses design criteria equally well. Students may champion their own ideas without adequately considering others' ideas or conducting objective analyses. We address this issue by asking students to collect both qualitative and quantitative data during testing. Students then reflect on the design criteria and assess how well each solution meets those criteria using the collected data. So, instead of a student's preconceptions, popularity, or perceived "smartness" influencing the group's design decisions, objective and impartial data play the key role.

Presenting the output of design testing as a number can provide a clear measure for students to compare one design to another. In some challenges, a simple test can provide a measure. For example, to test the strength of mortar in a stone wall, students use a "wrecking ball"—modeled by a golf ball on a string (Figure 5.5). They pull the ball back so the string is at a 10-degree angle and let it drop, then repeat with progressively larger angles, recording the angle at which the wall is destroyed. (Watch a video of this activity at eie.org/book/5d.)

However, quantifying data can be difficult given the tools and instruments that are typically available in elementary schools. Another solution is to ask students to rate a series of variables against a rubric with numbered scales. For example, as students assess the plant package they designed, they use a rubric to score how well their package functioned to contain, carry, display, protect,

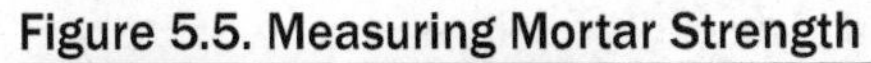
Figure 5.5. Measuring Mortar Strength

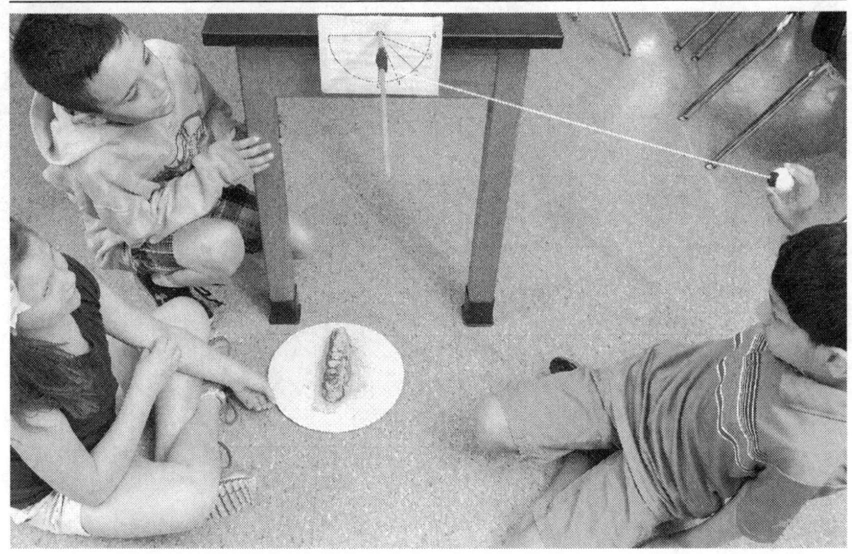

and preserve the plant, and also how well it communicated information about the needs of the plant. The sum of the numbers generated by their observations provides an overall "quantitative" measure of their design.

Whether measures are quantitative or qualitative, they must be sensibly linked to the goals and context of the design challenge. Although the units we design engage students in both quantitative and qualitative measurements, the main goal is for students to be able to identify what a successful design is and, with experience and scaffolding, develop and improve their own measures. In some of our units, students develop their own measurement rubrics by considering the desired outcomes of their technology and the ways they might measure them. (View a sample of our rubrics at eie.org/book/5e.)

Principle 7: Cultivate Collaboration and Teamwork. Collaboration is a highly valued skill in both science and engineering. Elementary educators are expected to help their young charges develop the ability to work productively in groups, yet many popular engineering programs involve contests. Competitive classroom environments that designate winners and losers can differentially discourage girls and students from racial and ethnic groups underrepresented in STEM. Many girls and some American cultures value interaction and collaboration more highly than competition (Lee, 2003). Collaboration affords students rich opportunities to develop expertise and identity as valued science and engineering contributors. Learning environments where such interactions are the norm allow students to be valued as contributors in a variety of ways. This reduces the impacts of race and socioeconomic status by allowing students multiple routes to social status instead of sorting and alienating them through hierarchies and competition (Carlone et al., 2011; Olitsky, Flohr, Gardner, & Billups, 2010).

Engineering lessons and tasks should be designed to foster collaboration and teamwork within and among groups. In almost every activity, students should have the opportunity to work with a partner or within a larger group. The lesson plans should make clear that comparing or rank ordering the solutions presented by different student groups is not a goal. Rather, each group compares its solution to a set of fixed design criteria and to prior designs that the group generated.

EiE teachers mention that learning to work in teams is an essential part of EiE lessons—it helps develop students' ability to work together and appreciate what their peers can contribute. A 2nd-grade teacher from Palm Springs, California, described how the teamwork and tactile nature of engineering encouraged a team of English learners to participate and communicate with one another, resulting in an outstanding design as well as building their confidence in their academic abilities:

> I had a wonderful experience this year in October in my classroom with the Building Bridges unit. My three shyest and lowest academic students, who are also all in speech for articulation and lack of language skills (one Hmong, one Cambodian, and one Spanish), ended up choosing one another to make up a group for the hands-on bridge design activity. Normally, these three would be in a group and would either not be understood or would be too shy to give their opinions or ideas. What I saw with these three working together gave me goosebumps of happiness! The three of them not only worked the *best* together, but they ended up making their prototype bridge before any of my other groups were even half finished. They offered one another ideas, and instead of overthinking or letting the "smartest kid" make all the group decisions, I saw them communicate in a whole new way, by just *doing*! They took ideas from our pre-lessons that even though they previously had trouble with getting their written ideas down on paper, I know they got something out of it as they used columns for support and attached paper clips and yarn as wires to make their bridge. They even thought to use straws as guardrails! After the Improve phase was over, this group had not only the safest, strongest, and longest bridge, but also the most creatively done bridge as well! It looked really cool and they experienced success early in our school year that they might normally not ever have had if not for this EiE unit. It gave them something that they were the best at, something in common to talk about, and they learned lots of new vocabulary as well as engineering concepts. I truly feel that this early success made a huge difference in these three

> students' future academic success, even *more* so than for my other students, as these three ended up making the most progress of anyone in my class. Thank you for designing activities that challenge our high[-performing] kids and let our lowest[-performing] kids shine!

Principle 8: Engage Students in Active, Hands-On, Inquiry-Based Engineering. Not surprisingly, students prefer hands-on activities and science experiments over text-based exercises. Such active learning experiences are still uncommon in many classrooms, but should be integral to engineering units. Inquiry-based instruction that emphasizes active student engagement in analyzing and making sense of data is clearly superior to techniques emphasizing passive student learning when it comes to increasing student conceptual understanding and inquiry ability (Minner, Levy, & Century, 2010). Various studies have shown that two groups in particular benefit from hands-on and inquiry-based learning: girls and African American students, particularly African American boys (Brotman & Moore, 2008; Kahle, Meece, & Scantlebury, 2000). One study found that middle school students who used a hands-on engineering design curriculum outperformed students who used a traditional science curriculum and students engaged in an inquiry-science unit on an assessment of science reasoning (Silk, Schunn, & Strand Cary, 2009). The evidence that hands-on work promotes engagement, interest, and learning should inform the development of engineering curricula. As one 5th-grader remarked, "I liked the way we worked in teams and actually built the circuit instead of just learning about it. I learned much more than I would have if I just read about it."

The merits of active, hands-on learning affect older students as well. Providing opportunities to access the material through many modalities remains a powerful experience for urban 4th-graders in Washington, D.C., as this teacher describes:

> Hands-on learning doesn't mean just touching manipulative cubes and counting. It really goes deeper than that. It allows you to feel material, measure material, build, create, problem-solve—lots of touch points, literally. I think for students who have trouble with attention, or trouble with sensory issues, or difficulty learning in traditional methods, this [engineering unit] really allows them to access knowledge in a different way. We made lots of touch points for that. They got to feel the materials. They got to drink hot chocolate out of cups to feel what the difference was [in the heat transfer] and then they did the scientific testing, which allowed for them to create data to support what they already felt and did and it had meaning. It's kind of multileveled. There's lots of touch points. It allows kids access to knowledge that they may not get through just reading it. It makes it more relevant once they felt it and manipulated the materials themselves.

A critical part of engineering for young students is the opportunity to physically manipulate their world as they understand how it works. This provides visual and often tactile experiences that can help them more deeply understand the concepts. One very talented kindergarten teacher in an inner-city school in Lawrence, Massachusetts, was convinced of the merits of engineering when she observed the impact it had on her students, particularly a few who had struggled throughout the year:

> This year, I had three children with physician-diagnosed ADHD. . . . They were constantly out of their seats, moving around every 5 to 10 minutes—until we started our EiE unit. I was shocked to observe that while they were engaged in the engineering challenges, all of them were focused and on task for up to 40-minute stretches. Furthermore, although ADHD children often have a difficult time with the writing process . . . I saw a huge investment and attention to task and detail with the writing components embedded in EiE. . . . It was remarkable! I think the combination of allowing the children to manipulate materials, to engage with them at their own level, to make cross-curricular connections, and to work on a challenge with a goal of finding a solution by trying again and again reached these students as nothing else that year had.

CATEGORY 3: SCAFFOLD STUDENT WORK

Students need teachers to guide them as they learn complex processes and transfer what they have learned to new problems. Instructional methods providing guidance to students who are learning actively have been shown to be consistently superior to pure "discovery" methods where students do not receive guidance (Hmelo-Silver, Duncan, & Chinn, 2007; Mayer, 2004). Scaffolds can include modeling and coaching to make a discipline's practices and ways of thinking explicit, providing expert hints and explanations, and providing structures and prompts to reduce cognitive load for students engaged in complex tasks. The following three design principles address scaffolding directly.

Principle 9: Model and Make Explicit the Practices of Engineering. Many disciplinary practices and learning tools, such as the engineering design process, are specific to engineering, as I've described in previous chapters. Providing explicit frameworks (as we do with our five-step process) and gradually removing teacher guidance can encourage students to take more initiative and responsibility over time (Cuevas, Lee, Hart, & Deaktor, 2005).

Students benefit from the explicit discussion and modeling of engineering practices. We see this in action as a 5th-grade teacher from Lexington, Massachusetts, guides her class through the construction of circuit diagrams in an electrical engineering activity designed to introduce them to standard symbols for schematic diagrams:

> ***Ms. King:*** What's the difference between the picture of the wire and the diagram of the wire?
> ***Leslie:*** One is squiggly, the other is straight.
> ***Ms. King:*** That's right; in diagrams, the wires are always straight, even though they're not in the real world. So, what symbols do we need to make the classroom diagram?
> ***Jared:*** A battery.
> ***Ms. King:*** Come up and get the symbol and put it on the board for us. What else do we need?
> ***Cate:*** A bulb. [Posts the symbol on the board next to the battery]
> ***Ms. King:*** If the wires need to be straight, then what will the corners look like?
> ***Tina:*** Right angles.
> ***Peter:*** Could it be a triangle?
> ***Ms. King:*** Do you mean like this? [Draws a triangle-shaped circuit with parts in it] It will never be that way. Why do you think so?
> ***Tom:*** So people don't wonder why it's that way.
> ***Ms. King:*** [Draws a square circuit] It always has perpendicular lines. Now, if I put a bulb here, a battery here—Tina, would you draw the wires?
> [Tina draws another square connecting to the first.]
> ***Ms. King:*** All right, I'm going to have you practice doing schematic diagrams now.
> ***Brian:*** Could you have something perpendicular in a different shape?
> ***Ms. King:*** Draw what you mean on the board, in a different color chalk.
> [Brian draws a circuit in an L-shape.]
> ***Ms. King:*** I don't think it's wrong, but you want it in the most clear way, like when you do fractions you want the simplest fraction. I'll give you papers now with a picture of a circuit at the top. Use the symbols and draw a schematic diagram.

In this excerpt, the teacher works with her students to establish the "ground rules" for making a circuit diagram, including the rule of thumb that it should be laid out in the "most clear" way. Circuit diagramming is a form of discourse practice in electrical engineering, and the guided discussion makes explicit the rules of this practice, a form of scaffolding that Nasir, Rosebery, Warren, and Lee (2006) call "making visible the structure of a domain" (p. 492). Aikenhead and Jegede (1999) labeled such teacher support "cultural border crossing." When we recognize that learning is a cultural process (not only a cognitive process) for all students and that all students need help to become comfortable with the culture of engineering, including its norms and practices, we are able to design learning environments that welcome everyone.

Explicitly describing the primary practice of engineering—the use of a structured design process to solve problems—helps orient students to what counts as engineering work and what is expected of them. The articulation of a process is a useful tool for students and allows all students access, as described by this Washington, D.C., teacher:

> I use the engineering design process in EiE units, but also in any other [engineering] units that I teach, so it's very much part of what kids do. While they may not be able to articulate each step specifically, they often are able to understand that there is a process. I think for kids who are coming to school with less background knowledge, less resources, this process allows for risk taking and mistakes and access to bits of information that are nearly the same for everyone. They come into these units on a level playing field based on background knowledge. Also by going through the process, it's something that all students can use in other parts of their lives. The idea that we want them to ask good questions and imagine possibilities and plan for those possibilities and create and test and improve. We want them to do that with everything—especially for the kids who may not have access to all of those things. Giving them a tool for a way to solve a problem is really critical because they may use that tool outside of this classroom, too, to help them succeed. I really believe that EiE helps underserved learners because it brings everybody in to a real-life situation that's new to everyone or almost everyone at the same time. Underserved populations tend to be missing background knowledge that's critical to problem solving a real-life, scientific, or engineering problem. With EiE, the problems are unique and interesting and diverse enough that it brings everyone to the same table at the same time. The process is such that by going through the EDP in the way that we do, all kids are able to get to the same end point regardless of what they came in with.

Principle 10: Assume No Previous Familiarity with Materials, Tasks, or Terminology. Children enter school with a wide range of experiences and backgrounds that can affect their achievement (González, Moll, & Amanti, 2006). Some children come from homes with rich oral or storytelling traditions. Some have exposure to many types of educational books, toys, or games. For some, material possessions abound. Some are exposed to a wealth of science experiences. Some have parents or relatives who are scientists and engineers. In an effort to minimize the effects of prior experience on students' success with engineering tasks, engineering lessons might include introductory activities that give students relevant experience with materials and terminology, allowing those who are English learners, for example, or those who have not previously encountered engineering materials to engage on a more level playing field. For instance, lessons should not assume that students know what a pom-pom is, or how index cards, printer paper, and wax paper are similar or different. Instead, an activity should permit students time to interact with the materials, label them, and explore their properties.

Principle 11: Produce Activities and Lessons That Are Flexible to the Needs and Abilities of Different Kinds of Learners. Elementary students differ greatly in their developmental and cognitive abilities. They learn best with a variety of modalities and supports. Fortunately, engineering challenges are well suited for differentiated learning. A simple challenge can be made increasingly complex by adding criteria and constraints. Engineering activities should be designed to accommodate a range of learners. Basic materials, for younger students or those with beginning literacy skills, may require less reading and writing and rely more on pictures than materials for older or more advanced students. More complicated challenges might ask the students to consider a larger number of criteria, constraints, and measures as they design. At both basic and advanced levels, tips for simplifying or expanding the activities and for scaffolding English learners can help experiences reach a broad and diverse range of leaders.

Many elementary classrooms contain students with a wide range of abilities. A 5th-grade teacher in Lauderhill, Florida, recounts how a well-designed engineering activity motivates and engages all learners:

> We have some of the students in our population who are either emotionally handicapped or they're labeled as a slower learner. What's nice about this curriculum is it allowed them to participate. There's so many aspects of the EiE curriculum—it can be hands-on, it can be an illustration, it could be a project. . . It's not pencil-and-paper. It's not just a textbook, and because of that certain students who are the slower learners are able to work with others.
>
> It brings them out of their shell where, "Okay, I know my weakness is reading." They can't read the storybook because it's already out of their grade level. But with EiE, there's other things we can do for them to understand it. We expose them to the vocabulary. We expose them with pictures, the illustrations that go along with the storybook.
>
> And then, guess what? You can take what you read about and heard, and now let's attempt to make it. Let's touch it, let's feel it, let's use your observation, your five senses to really dive into it. So, it takes them out of their shell, and a lot of our students who are lower readers participate the most. They are active, they're engaged, and they're actually collaborating with the people around them, the students, to work together. They might not be able to grasp and use the big vocabulary, but they can explain it to you. They can definitely explain to you what

went on—what the process was and "This is what I made." And they get so excited when they're able to be a part of everything, and not be separated.

I think it's a great feeling, because you see the students when they're done with any of the EiE lessons. They're knowledgeable, they have an idea of what they did. They can explain it to you, so they're more involved. I think they like that. A lot of times, maybe because certain students are labeled a certain way, we tend to leave them out in certain aspects. Or we don't include them in certain activities because "Well, they can't do it." Because they've been labeled a certain way. With EiE, everybody can do it. There's different levels.

CATEGORY 4: DEMONSTRATE THAT EVERYONE CAN ENGINEER

Students can enjoy "doing science" without wanting to be a scientist; their identity is shaped both by where they come from, in terms of place and in culture, and by their desire to have an effect on the world (Brown, Reveles & Kelly, 2005). Whether or not students wish to be engineers, we know students are more engaged, interested, and confident when they have both the responsibility and the opportunity to become more competent and to make choices about what forms of competence they wish to "specialize" in. The fourth and final category of design principles focuses on students' self-identification as engineers, or as being good at engineering and understanding the value of engineering.

Principle 12: Cultivate Learning Environments in Which All Students' Ideas and Contributions Have Value. Students are more likely to engage when they gain social capital through participation and when their contributions are valued. For example, Roth and Lee (2007) reported that a low-achieving student with learning disabilities became an "expert" when given the opportunity to engage in meaningful practices of environmental science; minority aboriginal students in the same project showed much higher levels of engagement when allowed to choose their mode of participation (Roth & Lee, 2004).

Environments where students are valued for their contributions are less likely to reinforce socioeconomic inequalities (or ethnic and cultural differences) and more likely to promote broad engagement. An equitable classroom is one where all students can negotiate and expand the roles they play, where all students share authority, where students are agents with ownership over their learning, and where all students have the opportunity and responsibility to participate with competence (Carlone et al., 2011).

Curricula that equitably engage all students must provide many different ways and opportunities for students to think strategically and collaborate productively with their peers. My team achieves this by carefully designing our engineering challenges. During our activities, students conduct experiments to learn how different materials or procedures perform, and then they decide as a group on a strategy for how to apply what they have learned.

We structure our lessons very purposefully to encourage all students to take responsibility for contributing to the group. For example, students brainstorm individually before sharing their ideas in a group discussion. The combination of complex challenges and varied participation structures create opportunities for students to engage, contribute, and share authority and expertise in the classroom.

Here's how one Washington, D.C., teacher reflected on this aspect of the curriculum after she completed a unit on engineering a solar oven:

> In this class, you may not know based on the video that there are actually a number of students who really struggle behaviorally and academically. They're not necessarily the same students. In this entire unit, we've had very little of that. They're fully engaged. Leading in their groups. Excited at the opportunities to show what they know. The pacing of the unit and how much thinking it takes, but again that there's not this right or wrong answer that there is safety in taking risk. If there are students who are really struggling academically, which there are many in this class that we have been teaching, they feel safe in taking risks, because they're learning something for the first time like everybody else.
>
> What I find is, sometimes those students understand the concepts faster than others and all of a sudden, they're revered in their group as the knowledgeable one—which may be a first for them.

Principle 13: Foster Students' Agency as Engineers. An equitable classroom is one where all students can learn, participate, and see how learning is relevant to their lives. Culture, gender, and socioeconomic status all play into students' views of themselves and what they think it means to be a "science person" in school (Carlone et al., 2011). Whether they are knowledgeable or struggling, students are more likely to adopt a science (or engineering) identity if they have the opportunity to be active and collaborative producers of knowledge for the classroom community (Kelly et al., 2017). These students will be more likely to affiliate with the discipline and seek out engineering opportunities in the future.

Engineering lessons can be designed to deliberately cast students as engineers using language that invites them to engineer solutions to problems. Teachers tell us that students quickly internalize their roles, excitedly proclaiming, "We are going to be mechanical engineers today!" or asking, "When do we get to engineer again?"

An exchange in Ms. Neat's 2nd-grade classroom in Hollywood, Florida, captures how teachers can help their students see their potential as engineers, and how students will readily adopt this identity (to watch, visit eie.org/book/5f):

Ms. Neat: At this point, do you guys feel like materials engineers?
Class: Yes.
Ms. Neat: Why? Anybody, yes.
Alejandro: Because we already, we already did like three [steps] of the engineering design process. And we asked a question and we answered, "What are solutions?"
Ms. Neat: Okay, anybody else? Do you feel like a materials engineer? And why? What makes you feel that way? Yes? Malcolm.
Malcolm: I do because I've been here all 3 days on our materials engineering process. I feel engaged and interested in what's probably gonna happen next. Just as a materials engineer would. And I think I'm really ready.
Ms. Neat: You are. Okay. Wonderful. I think you guys are all ready, too.

Later in this class, an African American girl articulated why she decided she wants to be an engineer. She stated, "I grow up, I would like to be an engineer, because I have been wanting to make things that would be helpful. When I grow up, I think I *can* make things that would be helpful for people." (Watch it at eie.org/book/5g.)

Principle 14: Develop Challenges That Require Low-Cost, Readily Available Materials. EiE engineering design challenges require low-cost materials that can be purchased at grocery, craft, or hardware stores for three reasons. First, students and teachers perceive engineering as more accessible when it calls for ordinary materials. Second, schools often have limited funds for materials. Third, when materials are easy to get, students who find school engineering engaging have the opportunity to continue their explorations outside of school. Many teachers we work with, including those who work with urban, underserved youth, report that their pupils continue to hone their solutions at home, after school or during vacations. One teacher whose class engineered solar ovens shared, "Some of my students who are normally hard to keep on track couldn't get enough of this activity . . . and were planning on creating their own solar ovens at home!" Another urban teacher learned that her students bought out all the flour at the local corner market over spring break as many of them continued to improve their playdough design at home. A Framingham, Massachusetts, 4th-grade teacher shared these two vignettes:

> My bilingual Spanish-English class was working on the Environmental Engineering unit, designing water filters to remove tea, soil, and cornstarch from water. One member of this class was a tall, rather awkward child of African descent from the Dominican Republic. She seemed to lack confidence and had never fully engaged in many academic activities. When we began working on water filtration, she

commented that this was a big problem in her home country, and she worked well with her classmates to clean the water during the design challenge. At the end of the unit, she asked to take home a copy of all of her group's designs. I thought nothing of it and gave her the papers. Several months after we completed the challenge, she came into class and said, "Señora N., *lo hice*, I did it! I made a filter that gets the water perfectly clear. Could I stay after school and show you my designs?" (I should note that it is really difficult to remove all tea color from water.) This child had returned home, and because she could readily find the materials for the design challenge (she literally cut up old T-shirts and towels), she had continued to work on her filter design for months! I let her show the class her final design. I discussed her interest with her mother, a very supportive woman with limited formal education but a desire to see her daughter triumph in this country. Shortly after, my student told me, "For my next challenge, my mother has arranged that when I return to the Dominican Republic this summer, I am going to visit a man at a water treatment plant who is working on the desalinization of ocean water for drinking water." The EiE challenge had piqued this child's interest, and she connected it with a problem she knew affected her family's homeland. This child is now in the gifted-and-talented program and continues to thrive in math and science at the high school.

Last year, I worked with a very challenging student population; many of my students had been exposed to trauma and lacked trust in adults and one another. This was often manifested through a lack of motivation on school assignments, as well as outbursts of anger and defiance. That all changed when my class began working on the EiE chemical engineering unit, in which students design a process to make playdough. My students became so motivated that almost half of the class brought in playdough samples and processes that they had developed at home to try to improve the processes they were working on in class. During class, they would discuss the quality of their samples and try to figure out ways to combine their ideas to come up with the best possible sample. I had to extend the unit for several days as they begged me for "more engineering time." We had a comparison of wheat flour versus corn flour, as many of my Mexican and Central American students had access to corn flour at home, which brought a cultural dimension to the activities that gave them pride in their identities. By using only materials that even my students of most limited resources had at home, the unit made the concepts accessible to all students. Several students ended the unit saying that they wanted to become engineers as adults, because that was something they knew that they had the ability to do.

Well-designed curricular materials are critical for attracting, engaging, and retaining students' interest in and positive attitudes about engineering. I share these principles to invite anyone involved in resource design or classroom instruction to use them as a starting point for conversations about what high-quality engineering materials should look like. I encourage others involved in K–12 engineering programming and resource development to carefully consider other ways to actively reach out to underrepresented populations—for unless we do, the status quo in which certain groups are underrepresented will continue.

CHAPTER 6

Teaching Engineering

If you're an educator, you know introducing a new discipline into any school or classroom is a herculean task. Throughout the previous chapters, I examined the most common questions related to implementing an elementary engineering curriculum—from the philosophical "Why should I teach engineering to my students?" to the curricular "What does engineering look like at the elementary level?" to the pedagogical "How can engineering engage all of my elementary students?" All of these questions might leave you still wondering: "How can I teach engineering to my students? What are the challenges and concerns I will face when I attempt to teach engineering for the first time?"

My team and I have supported teachers nationwide as they transitioned from novice to expert—from when they first start teaching engineering to when they become confident elementary engineering educators. In this chapter, I write specifically to teachers and educators. I explore the top concerns that teachers have as they begin to teach engineering and I offer effective strategies, which I hope will make your job easier and help you and your students succeed.

TEACHER 1: "I DON'T HAVE ANY EXTRA TIME!"

Frequently, teachers lament that they don't have time to teach engineering in an already-packed school day. Unfortunately, we can't create more hours in the school day, so to better support teachers, my team and I thoughtfully considered the ways in which engineering design challenges can support existing instructional goals. We knew there would great opportunities for engineering to reinforce the science and math concepts that elementary students learn throughout the year. For example, students who are challenged to design a package for a plant that will keep it alive need to consider the basic needs and structures of plants so their design can accommodate these. Research has shown that when engineering activities are specifically designed to reinforce science concepts, students learn the science better (Lachapelle, Oh, & Cunningham, 2017) (see Chapter 7). Students understand concepts more deeply when they are able to apply their abstract knowledge to specific hands-on tasks. Engineering lends itself well to project-based learning. It provides an overarching problem that students solve by drawing upon knowledge gained from many disciplines. As you teach engineering, you can reinforce topics like science and math, but also literacy and social studies. For example, when students are asked to determine where to site a TarPul bridge for a community in Nepal, they learn that they need to consider societal and cultural factors, such as where the villagers want the bridge to be, as well as technical considerations related to the geography. In a culminating activity, they write a letter to the village elders to recommend a location and communicate how they reached their conclusion, thus drawing on language art skills such as persuasive writing.

If you're thinking about how to carve out a space to teach engineering during your school year, start by looking at the scope and sequence maps for the science topics you teach. Then choose or create engineering instructional units that align with that content. As you do so, consider how an engineering-themed lesson can reinforce the skills and ideas your students will learn in math, English language arts, or social studies. In this way, you're not only making time for a completely new subject, but you're helping your students understand other subject areas more deeply and effectively.

Of course, integration and connections can help alleviate time pressures, but it's true that engineering challenges will also require some dedicated classroom time, as there are new concepts to discuss and design challenges to do. Some teachers with highly prescribed curricula choose to offer engineering during times of the year when the constraints are more relaxed—for example, after state testing occurs or during time periods when student energy levels are high and are best harnessed by active units, such as the week before school vacations. Regardless of when and how teachers integrate engineering, most find that the skills their students develop are worth the time investment.

TEACHER 2: "THERE ARE SO MANY MATERIALS TO PREPARE!"

Before students tackle an engineering design challenge and build their solutions, they need to be able to handle and manipulate the base materials. Luckily, students, particularly young students, are very excited for the opportunity to touch and use different materials as they learn about the materials' properties and brainstorm potential design solutions. However, using materials in the classroom can present challenges for teachers.

To make your job easier, make sure you prepare, prepare, prepare. You can free up time later on (and reduce in-classroom headaches) if you prepare and make sure all necessary materials are prepped and ready to use. For example, if students will need a 4-inch x 4-inch aluminum foil sheet, cut these foil squares ahead of time. Create a list of all the necessary materials and gather everything that is needed before the unit begins. Count out the materials, cut the materials (when necessary), group sets of materials together, and compile the prepped materials into separate and organized bags or boxes. To help students focus, we have found that handing out small "materials bags" that contain one sample or a small 1-inch square can allow students to touch and explore the materials while keeping the focus on the material and not, for example, on how long the aluminum foil is when it is uncoiled from its roll.

Some design challenges may use a base model as well. For example, students might be challenged to design only one part of a more complicated system—blades for a windmill, a maglev train for a track, the insulation for a solar oven. If you teach a unit that requires a base model, build the model ahead of time. After you complete the unit, save the base model (if you have space) to use year after year.

If possible, we recommend recruiting materials helpers, too. Colleagues, parents, student helpers, or even corporate volunteers can all assist in assembling the necessary components for your students' engineering challenge. All this prep work will make distributing and managing those materials during class much easier.

During the lesson, your students will be eager to touch and manipulate these materials. Because the materials are new and fun, it's important to give students time to feel and explore them. Structuring time for students to consider the materials' properties and become familiar with the materials will inform how students use these materials later in the design challenge. Just make sure to communicate important instructions before students receive materials to better ensure that you have their full attention.

When students begin to create their own, unique designs, each group may use different materials in different amounts. Thus, you will want to consider the classroom layout and how you arrange the materials. Students will stand up and move around as they procure the items they need. Many teachers recommend using a "materials station" or "materials store." The teacher, an aide, or an adult classroom helper sets up one location that houses all materials and then oversees material distribution from that spot. After groups have developed an idea and sketched a design, they list the materials it requires. One member brings the materials list to the store manager, who doles out the specified items. Having a "materials station" prompts students to create a plan and then adhere to that plan as they build. This process not only mirrors a real engineering process, but it also helps prevent students from becoming too distracted by all the options. Instead, they focus on the purpose of their technology and what they need to design it.

At the end of the challenge, ask students to return any unused materials, deconstruct unwanted versions, salvage materials that can be reused, and help organize materials for later use.

TEACHER 3: "HOW AM I GOING TO MANAGE MY STUDENTS' VARIED, CREATIVE DESIGNS?"

Managing a classroom in which students tackle open-ended engineering challenges that allow multiple solutions may seem daunting to you. You may be wondering how you will keep track of each group's unique design and provide meaningful feedback. Teachers tell us that students embrace the opportunity to create an original design. However, they also tout the value of establishing some parameters. For example, teachers have suggested the importance of limiting the materials that students might use, at least initially. Having students choose from specified materials (instead of using a wide array or whatever they bring to class) focuses them and their work. It allows them to explore these materials in more depth to understand their properties and how they function. The common materials also provide a base that allows students to share and discuss what they have learned with one another.

Groups learn from one another when they focus on a common challenge. If all students in a class tackle the same problem with similar materials, you can become familiar with the types of designs student groups generate and the problems they might encounter. Working with the class to identify the criteria and constraints of their engineering problem helps you tailor your comments and feedback to what your students will need to achieve their goals. For example, in a class working to design a process for cleaning up a model oil spill, students are told that one constraint they face is in the materials available—they can choose from six different types of materials, each which has a price. As a class, you have discussed the goal: to create the cheapest solution possible that also removes the maximum amount of oil from the water and the shoreline. Your experience leading this challenge in previous years has taught you that students need to be reminded that each time they dip a material in the oil, they need to log its price. You also know that they will be surprised at first, and then frustrated, by how much it costs to clean a spill and how much oil still remains on the water even after their process is complete. They will need encouragement and questions that help them focus them on improving their processes when they begin the redesign phase.

As student groups explore possible design solutions, you'll guide them through this process, modeling and asking the kinds of questions that develop critical thinking and reflection (e.g., "Let's see what happens if . . .", "I wonder . . .") or the data-driven actions and decisions that should guide their work. Your questions will guide students to think about their technology and its design more deeply. In some cases, you might encourage students to focus on the design challenge's goal or focus their attention on an element of their design that you suspect will be problematic. Over time, you'll develop a trove of general engineering questions, as well as questions that are tailored to a specific design challenge and that will focus student attention on relevant aspects of the problem. Some common engineering questions include these:

- How does your design reflect what you learned yesterday/in previous lessons?
- Why did you choose these materials?
- What did you notice when you tested your design?
- Did your design work (as you expected)? Why or why not?
- How did the design fail? Which aspect(s) of your design might be the problem?
- How might you improve your design? How can you redesign your technology to [address the goal better—make it faster, cheaper, stronger, etc.]?
- Did this design work better than the previous design? Why do you think this is the case?

In addition to walking around the classroom and providing feedback to individual groups, it is helpful and important to structure time for whole-class reflection and discussion so that students can learn from their peers' efforts. Periodic whole-class discussions allow student groups to share their current designs. You can ask students to describe similarities and differences across the various solutions and encourage them to consider the general principles that underlie the challenge. For example, during a group discussion on how to design an effective sail, students may begin to consider whether bigger sails work better because they capture more wind. You want to encourage and remind them of the general science principles that they already know—that is, big sails aren't always good for a boat's design because sails that are too large can capsize a boat. Asking students to share their data, designs, and ideas may spark new ideas for their group's next redesign. More importantly, this activity can prompt students to reflect upon the larger engineering and science principles that inform their design challenge. As students present their "final" designs in a showcase or test their solutions in front of the class, they should share information and explanations, as they reflect upon their work and celebrate their learning and successes as a class.

TEACHER 4: "MY STUDENTS WON'T LIKE IT WHEN THEIR DESIGNS DON'T WORK."

Many teachers worry about their students feeling discouraged or upset if or when their designs fail. But this behavior or reaction stems from a belief that failure is an intrinsic or personal trait—and it's not. Failure is an extrinsic trait that should only be attributed to a particular design or to a group's inability to work together. We encourage teachers to recast the role of "failure" during engineering challenges (and in the classroom more generally) by emphasizing that engineers "fail often to succeed sooner" (Nightline, 1999). As you lead classroom discussions and presentations, scaffold your students' understanding of failure. Remind your students that failure happens often during the engineering design process and that all engineers face failure. When failure occurs—and it will—reinforce the idea that only the design has failed, not your students themselves. Communicate and model that there's much to *learn* from failure: What did not work this time, why not, what can they do to improve their next design? The answers to these questions can inform the next design iteration.

We find that it's also helpful to downplay competition between individual or group designs, particularly in the early grades. We suggest that students focus more on working as a *class* to generate the best design solution possible. In this classroom dynamic, successes and failures are shared by all. This better enables all students to celebrate original ideas, the effective parts of each design, and overall high-performing designs. As students engineer in this dynamic, they often come to recognize the strengths their peers bring to a project as they reconsider traditional classroom norms and hierarchies.

It will take students time to adjust, accept, and even welcome failure when it happens. For example, some teachers report that their high-achieving students, many of whom may not have encountered failure before, have difficulty facing failure. However, these teachers also mention how much they appreciate the opportunity to have an open conversation about failure, encouraging their high-achieving students to reconsider their relationship to failure.

TEACHER 5: "I DON'T KNOW HOW TO GRADE MY STUDENTS ON ENGINEERING."

Our goal as educators is to nurture students' ability to tackle open-ended engineering challenges. As they do so, we strive to develop their engineering habits of mind. The assessments we use need to reflect these goals. We need to communicate clearly and consistently the expected outcomes. For each engineering challenge we create, we specify criteria and constraints to guide students' work.

Assessing the *product* is only one part of an engineering project. What is perhaps more important, particularly for elementary students, is the *process* of engineering. As students engineer, they should become more comfortable asking questions, finding answers to them, creatively brainstorming, considering others' ideas, handling physical materials to create a prototype, gathering data, making decisions based on data, collaborating with their teammates, and communicating what they have done and why. These are just a few of the desired practices and habits. As you provide feedback to your students, keep in mind the practices and habits that you would like your students to develop. The feedback and evaluation that students receive from their educators signal what is important. How they work to generate a design solution and what they do to further improve the solution should be more significant than any one particular design.

You may find it helpful to assess students' engineering efforts with a combination of a portfolio and ongoing performance assessment. Through our work with educators, we have found that scaffolding students' work with a semi-structured portfolio—a set of worksheets or a student notebook—prompts students to document their ideas and reflect upon their work and data while providing a cumulative log for students and teachers. Students' daily work in the classroom offers an ongoing performance assessment. Periodic whole-class "shareouts" provide more formal opportunities for students to explain and present their current work. If ongoing documentation of students' progress is required, in addition to engineering journals and portfolios, we recommend creating rubrics to document each child's work, performance, and contribution to the activity. In our curricula, we articulate learning objectives for each lesson and then develop rubrics for assessing them. Such tools provide concrete referents for evaluation if they are needed. The critical element is that the assessment tools align with the types of work and learning you want to foster in your students.

TEACHER 6: "I HAVE KIDS IN MY CLASS WHO WILL BE LEFT BEHIND."

Engineering challenges, if thoughtfully designed, can provide a range of entry points for students. Students can tackle a simple problem with only one constraint, or the same problem with a number of specified constraints and criteria. Challenges might "hook" students through an interesting context, the relevance of the problem to their lives, or a role model they admire. Problem solving might entail little reading or documentation or it might require careful data collection and analysis. The opportunity to move around the classroom, manipulate physical artifacts, and think "outside of the box" can draw students into the task. Engineering activities can engage a wide spectrum of students, such as English learners, students with special needs, students who do not best express themselves in writing, active students, artistic students, and many others. Engineering invites students to "show what they know" by demonstrating, often with a physical model, their solution to a problem. Students are able to pursue creative ideas they find interesting and learn from failed attempts.

Tailor your challenges so that they are age-appropriate and accessible to all students in your classroom. But don't presume too much about students' capabilities ahead of time—they may surprise you with what they are capable of when challenges are offered in this way. As students get older and gain experience, they become more facile engineers who can address increasingly complex problems. As with engineering in the real world, design problems become increasingly difficult as students consider additional constraints and criteria. You can set forth an initial engineering problem, and as groups successfully meet the original goals, you can challenge them further by introducing more and more difficult specifications to meet. For example, the basic challenge may ask students to design a water filter that removes particles from dirty water. Once successful, a group might also be challenged to optimize the cost of their filter—the less it costs, the better. The renewed challenge requires the group to rethink which materials they use and in what quantities. An even more advanced version would ask students to design a filter that removes both loose particles and color from dirty water. Within any class, after students successfully design for the initial parameters, they can move on to more challenging scenarios.

As in all pedagogy, developing fluency in engineering instruction takes practice and reflection. Teachers report that they begin to feel comfortable the third time they lead an engineering activity and start to feel like experts around year 6. We have worked closely with teachers to develop resources that can support them as they implement engineering with students. Professional development workshops (either face-to-face or online) foster familiarity with activities and provide philosophical and instructional structures. Our differentiated resources collection provides additional supports for students such as English learners or students who receive special education services. We've created rubrics and assessments for teachers to use. However, the resource most requested by teachers has been video from real classrooms doing engineering. Hearing this plea for the past 4 years, we have created an elementary engineering video library (see eie.org/video). It contains footage from more than 50 elementary classrooms engaged in engineering design challenges. Watching practicing teachers from across the country enact engineering with their students offers powerful models for guiding students in active problem solving. As the teachers teach engineering with growing confidence, the credibility of the approach gains traction in aspiring teachers' minds.

Part III

LOOKING TOWARD THE FUTURE

CHAPTER 7

Impacts of Engineering Education

When I started the EiE project, I knew I had a lot to learn. Elementary engineering education was new terrain—my previous focus had been on middle school, high school, and college-level science and engineering. To develop effective materials, I would need to get acquainted with the capabilities of young children and refresh my understandings of the day-to-day practices of elementary classrooms. I recognized that I would also need to get smarter about supports elementary teachers valued and messages about elementary engineering that resonated with administrators. In the first chapter, I outlined a number of reasons for engaging elementary students in engineering. But how would we know we were achieving our goals and aspirations?

From the start, I was committed to a data-driven curriculum development process. By this, I mean that my team and I would always make decisions that were based in data. Unfortunately, when I started in 2002, virtually no data or educational research about elementary engineering existed. There was research in related fields such as science and technology education. There was research on access and equity in education as well, which I knew would be useful. But because so little research existed back in 2002, we had to develop and conduct our own research to support our materials and our claims.

We adopted a "backwards design" approach (Wiggins & McTighe, 1998). We ask ourselves what we want students to know and be able to do after engaging with our curricula. Once we feel confident that we understand what we want students to achieve, we focus on creating materials and experiences to support such outcomes. Then we create research and evaluation instruments to collect information from students and teachers. Always, we look at the individual pieces of data and compile them so we can draw out trends in the data.

I made the decision early on that our goal was not just to produce the best materials possible. Our goal would also be to understand how students and teachers learn engineering and to contribute to knowledge in this new field. We created research and evaluation instruments to collect crucial information from students and teachers. For example, to understand what students think engineering and technology are, we created instruments that ask them about these topics. We asked hundreds of teachers, and ourselves, what students should know and do in elementary engineering. We asked questions about whether our resources worked for teachers in their classrooms. We asked whether students were engaging in engineering and learning what we hoped they would. We asked whether our resources were reaching all students. Teachers' feedback has been particularly important in the design of our materials.

Additionally, I knew it would be challenging to convince teachers, schools, and districts to allocate resources—the most important of these being time and funding—to another discipline, especially one that was not "core" and, at the time, was not present in most state standards. To convince administrators that engineering was worth the investment, I recognized that we would need to provide statistical data demonstrating its impacts on students' learning.

Of course, all this was easier said than done. First, we would need to figure out what elementary engineering should entail as we simultaneously tried to measure its impact on students—meaning that we were measuring a constantly evolving curriculum in development. Second, we guessed that many of the most important features of engineering (such as persistence through failure) could not be captured by bubble-scan tests. Third, we would be assessing these outcomes before we could document what engineering looks like in elementary classrooms and before we really knew what was important to assess. However, to get school administrators to consider using our materials, we would need to develop metrics that utilized a quantitative approach, which meant focusing on a few outcomes that we could measure using statistics. We would need to concurrently figure out on the fly what students were capable of and how to support their learning as we also tried to understand the benefits of engineering and how to measure them.

Over the years, my team and I, as well as a number of external researchers and evaluators, have studied the implementation and impact of engineering education and our EiE resources. To date, we have collected more than 375,000 surveys and research instruments, and we have spent thousands of hours observing classroom engineering. We've compiled the research and evaluation reports, papers, chapters, and articles about EiE—they now number more than 100. (You can find the research instruments and papers at eie.org/research.) I've been fortunate to work alongside passionate researchers who all took the task of creating an elementary engineering curriculum as seriously as I did. In this chapter, I summarize what we have learned so far about the impacts of engineering education on elementary school students, teachers, and schools. I focus on studies of EiE because currently there are very few other classroom-based research studies of elementary engineering.

IMPACTS ON STUDENTS

Students Who Use EiE Learn About Technology and Engineering

What were students' ideas about technology and engineering before and after they completed an engineering unit? My team and I wondered how participation in classroom engineering could alter students' understandings and appreciation of technology and engineering. We suspected that students would develop a more accurate and robust understanding of both, but we wanted data to back up our claim. To get those data, we needed to develop research instruments that would probe students' concepts of engineering and technology as I describe in Chapter 1.

Using these instruments, we found that our engineering curriculum has a dramatic, significant impact on broadening students' understandings of technology. A number of studies using control groups reinforced the fact that students gain a more accurate and nuanced understanding of technology after engaging in engineering. As Figure 7.1 shows, after completing an EiE unit, students are much more likely to indicate that commonplace, simple technologies, such as brooms, baskets, and bicycles, are technologies. In response to the open-ended question "How do you know if something is technology?" students are more likely to answer that technologies are human-made, and that technologies are designed to solve problems.

Figure 7.1. Change in Students' Responses to *What Is Technology?*

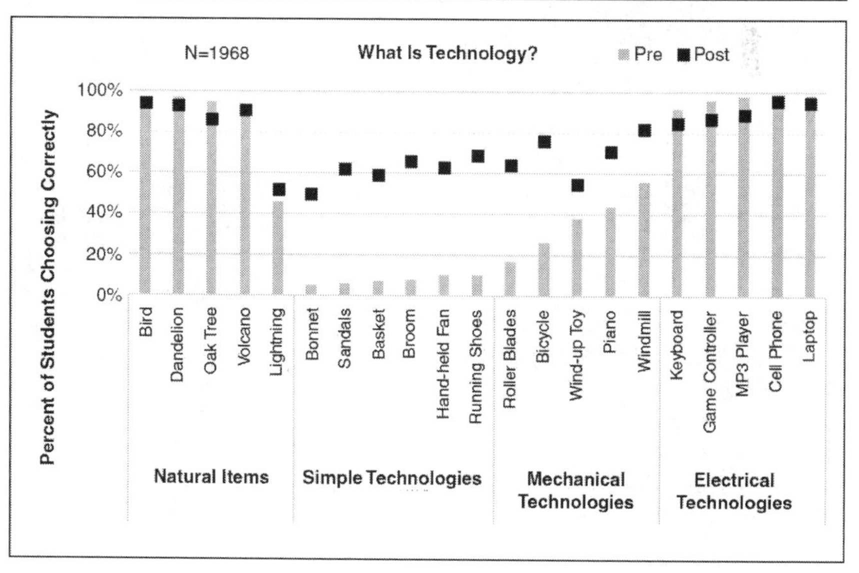

Figure 7.2. Change in Students' Responses to *What Is Engineering?*

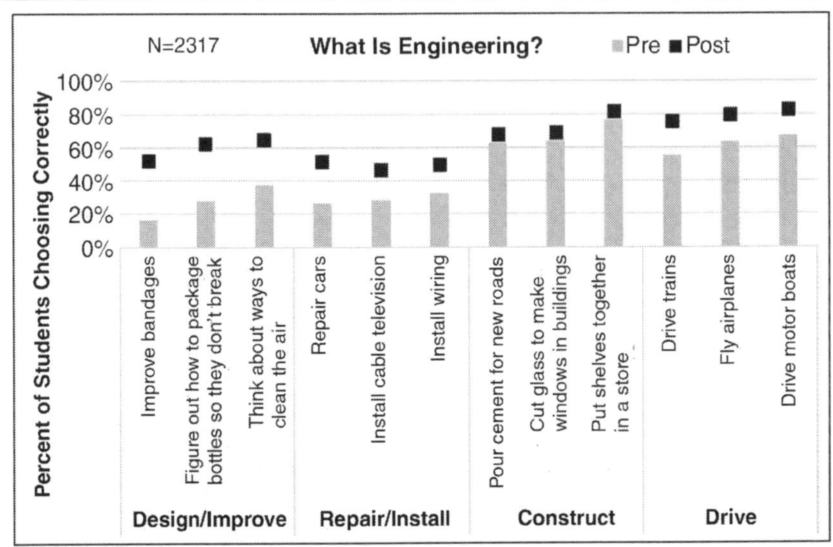

When students engage in engineering, they also develop more accurate understandings of the work that engineers do. We asked students what they thought engineers do for their work using our *What Is Engineering?* instrument before and after using EiE. As described in Chapter 1, before students engineered, they focused on the *object* of the task to select their responses. For example, they typically chose responses that included the word *engine*, *computer*, or *machine*. After engaging in an engineering unit, students were more likely to focus on the *type* of work—people designing, improving, testing ideas, and so on. Figure 7.2 shows that after engineering instruction, students are more likely to recognize that engineers design or improve objects. They more frequently select items such as "improve bandages," "think about ways to clean air," and "figure out how to package bottles so they don't break," which are some of the items we use to calculate a student's "Design/Improve" score.

Also of note is that after engineering, students are more likely to let go of their incorrect ideas that engineers "Drive" machines, "Construct" things, or "Repair/Install" things (also shown in Figure 7.2). For example, when asked to describe the work of engineers, students are less likely to select: "Fly airplanes" (used to calculate the "Drive" score), "Pour cement for new roads" (part of the "Construct" score), and "Install cable television" (part of the "Repair/Install" score) after they engage in EiE engineering activities.

Students Learn Basic Engineering Concepts from Engineering Curricula

We developed our curriculum so it would be age-appropriate. As students engage in engineering activities, the basic engineering concepts they need to learn are simple. We thought carefully about which concepts were important to include and how to simplify engineering concepts while keeping them accurate. For example, in our "Designing Bridges" unit (EiE, 2011e), students learn about piers, abutments, and the span of a bridge. Activities introduce them to some basic types of bridges and some of their advantages and disadvantages (for example, beam bridges are inexpensive and easy to construct, but can't span far; suspension bridges are more complex but can span a long way). Students learn about loads on bridges, and they learn about the kind of work civil engineers might do. In our "Designing Plant Packages" unit (EiE, 2011f), students learn about the functions of packages: to contain, protect, preserve, dispense, and display the product; to communicate to the consumer; and to make it easier to carry the product. They learn about package engineering as well, and they learn to do a simple needs analysis, matching the needs of products and consumers to the design of a package.

Figuring out how to reliably measure what students learn about engineering concepts was challenging because of a tension between our core commitments and measurement. We believe engineering education should primarily develop engineering practices and habits. But these attributes are very difficult to measure with the types of bubble-scan instruments that are used in large-scale research studies. We needed to develop assessments that did not distill engineering to a set of factoids. After a number of vigorous discussions, we created instruments that asked students to reflect on the processes of engineering, to consider the properties of materials in making design decisions, to apply fundamental engineering terminology, and to demonstrate their grasp of the types of work that engineers in a specific field might do.

Some of our research exploring what students learned about engineering compared students who studied both science and engineering to a control group of students who studied science concepts but did not do engineering. Other research we conducted compared two different types of elementary engineering instruction—EiE was compared to an engineering curriculum that had the same learning objectives but did not include our inclusive principles. After reviewing the pre- and post-assessments of all cases, we found that students from all demographic groups—boys and girls, students of all races and ethnicities—who engaged in an engineering activity demonstrated on the posttests that they knew significantly more about engineering concepts than they did before instruction.

How did engineering activities improve students' understanding of engineering concepts? Many students showed they had a better grasp of relevant engineering and technology vocabulary, and they demonstrated a basic grasp of the specific field of engineering they studied (for example, mechanical, chemical, or biomedical engineering). Students also showed a better understanding of materials, their properties, and their uses in engineering design scenarios. As I explained in Chapter 3, we consider understandings of properties and uses of materials to be a key goal of elementary engineering. Not surprisingly, students also showed a better comprehension of how to improve the specific technologies they had studied, such as bridges or packages. More interestingly, many students showed that they could use criteria to judge the effectiveness of a technology—an important critical thinking skill.

Students Who Use EiE Improve in Their Understanding of Key Science Concepts

We designed EiE specifically to help students apply science concepts as they engage in engineering design. For example, in the "Designing Plant Packages" unit, students think about plant health and survival needs as they design a package that will keep a plant alive. In the "Evaluating a Landscape" (EiE, 2011g) unit, students need to understand that rivers tend to erode at bends and curves, and how the makeup of the soil in a particular location can support a foundation—or not—so they can make decisions about where to place the foundation poles for a TarPul bridge. Asking students to draw on their science knowledge to solve problems mirrors what occurs in professional engineering. But this integration was also important because if we could demonstrate that engineering bolsters science understanding, teachers would have additional leverage when they argued to include engineering in their school curriculum.

As mentioned previously, our units all had both engineering and science objectives. In addition to developing curricular materials, we spent a lot of time creating assessments that would help us understand if students were making progress toward those objectives. We generated many questions and tested the effectiveness of these questions by asking students how they understood the questions we asked of them. We sat one on one with students and asked them to "think aloud" as they completed the questions, explaining why they chose a certain response. By doing this, we were able to follow the students' trains of thought and evaluate whether our questions were effective in the way we hoped. Such testing helped us make better instruments with strong evidence for validity. For example, in the "Designing Plant Packages" unit, one of the science questions asked students to select which response choice was true about plants. One of the options (the correct one) was initially worded, "Leaves make food for the plant." One student we observed chose this option and explained his response: "People and animals like to eat leaves, like lettuce." Although his selection was correct, the reasoning was not aligned with our intentions! Such insights allowed us to rephrase poorly worded options to be less ambiguous.

After such detailed testing, we created pre- and post-assessments of science understanding for each of our units. We compared students doing science and EiE to those doing only science. As part of our study of inclusive principles in engineering curricula, we also compared outcomes of students using EiE to those doing engineering units that don't explicitly incorporate science concepts. What we've found is consistent, even with our most rigorously designed large-scale studies: Designing engineering lessons so they connect with science concepts helps students learn both subjects effectively. Elementary students have a stronger grasp of the target science concepts after they participate in an EiE unit than an engineering unit without explicit connections to science. When students apply and use science concepts during a hands-on activity, they learn the science concepts better and more deeply. This is good news for teachers who wonder whether they should integrate another subject (engineering) into their packed school day. This finding suggests that as students engage in carefully designed engineering activities, they will not only learn about engineering but will also solidify their science knowledge.

Students Who Do Engineering Become More Interested in the Discipline

As curriculum designers, we wanted to know what kinds of engineering work would interest students and how they would feel about engineers and their impact on the world after completing our units. We were particularly interested in differences between girls and boys and between different races and ethnicities, as we seek to understand and effect change in the current makeup of the engineering profession, which is overwhelmingly White and male.

To measure students' perceptions of engineering, we developed another research instrument, the Engineering Interests and Attitudes survey. Our earliest version contained 20 statements about students' interest in various types of engineering jobs and work, and their attitudes about scientists and engineering—for example, "I would like a job where I could invent things" and "Engineers help make people's lives better." Students were asked to rate each statement on a scale from "Strongly Disagree" to "Strongly Agree." Later, to gather more information, we expanded this survey with items we drew from science interest and attitude surveys published in the literature, which we modified to reflect engineering. These new items addressed gender bias, the value of engineering, and students' enjoyment of engineering—for example, "Boys are better at engineering than girls," "Engineering is really important for my country," and "Engineering is fun." We also asked kids to rate their agreement at points in time: "Last summer I would say" and "Now I would say." We've used this survey with students in grade 3 through middle school both in classrooms and in out-of-school (after-school and summer camp) settings.

We've found that engineering education does indeed have a positive impact on students' interest in and attitudes toward engineering. Students of all demographics became more interested in and positive about engineering after participating in an engineering unit. After engaging in engineering activities, students are more likely to express that they enjoy engineering, and more likely to say they want to learn and do more engineering in the future. They are more likely to see value in engineering for people and the world. They are also *less* likely to agree that engineering is a field where males are more able than females.

Interestingly, our research aligned with previous studies by other researchers that indicate girls and boys are interested in different sorts of jobs. Boys tend to show more interest than girls in questions that have to do with inventing, figuring things out, cars, and structures. Girls, on the other hand, are more likely to show interest in engineering jobs that help society and people, such as developing new medical devices or ways to protect the environment. Both boys and girls show increased interest and more positive attitudes toward engineering and science, but girls show even larger jumps in interest and attitudes than boys. Though boys start out more interested in engineering, girls nearly catch up in level of interest after doing an EiE unit. As one teacher told us, "The strongest impact was with my female students, who were still talking about the engineering unit at the end of the year. They said it was their favorite thing they had done in 1st grade, and we did a lot of neat stuff!"

Students from Groups Underrepresented in STEM Show Increased Interest and Achievement

As we did the studies described above, we also investigated whether and how engineering outcomes differed by various populations of students. In both our pilot studies and our carefully designed large-scale study, we found that students who are underrepresented in engineering learn engineering and science concepts when they do engineering using EiE. Students who were English learners, were from a low-income family, or were Black or Hispanic tended to have lower scores on the science and engineering pre- and post-assessments that we administered (described in the above sections), but, importantly, they improved just as much as students who were not from these groups. During engineering lessons, all students were learning at the same rate. In some cases, participation in EiE helped close the achievement gap between Black and Hispanic students and White students, and between students with lower initial science achievement and those with higher initial achievement. These findings did not apply for girls—their pre-assessment and post-assessment scores mirrored those of boys on science and engineering assessments.

We were interested in whether teachers noticed any differences in students, and they did. We've done a number of surveys asking teachers how their students from different demographic groups engage with engineering activities. Most teachers report that all students are positively affected by EiE lessons, whether they are academically low or high achieving, including students with cognitive challenges, linguistic challenges, behavioral challenges, who are gifted and talented, who are girls, and who are Black or Hispanic. One of our earliest surveys found that teachers agreed most strongly that the curriculum positively affected the interest and engagement for students from races and ethnicities underrepresented in engineering (Faux, 2008).

A later survey found particularly strong impacts on both girls and students from races and ethnicities underrepresented in engineering, not only on students' interest in engineering, but also their interest in science and mathematics (Weis & Banilower, 2010). One teacher reported, "I see the same engagement level in EiE activities with underrepresented populations as I see with all my other students. I am not seeing the gaps that I see in other subjects. Underrepresented students are as capable in EiE activities as all my other students." A repeat of the survey during our large-scale study found the same results.

ENGINEERING IN CLASSROOMS

To understand more deeply how students experience engineering in classrooms, EiE researchers conduct qualitative research studies that draw from classroom observations, videos of students and teachers engineering, interviews, and student work. In other words, we want to move beyond variables that can be captured by standardized assessments and unpack the intricacies and nuances of how teachers teach and how students learn in classroom settings. Our team has an internal group dedicated to research and evaluation. We also are fortunate that other researchers across the country have chosen to study elementary engineering using the EiE program. I describe briefly some of the findings of these studies in this section. As we understand more about how engineering is taught and learned, we draw upon the results of these studies to enrich our program.

We wondered early on how teachers used our curriculum and how students experienced engineering. What messages about engineering did teachers communicate as they structured activities and led discussions? We closely examined footage of a teacher's classroom as she taught one EiE unit to investigate how she used the curriculum and specific teaching practices to support learning opportunities for her students (Cunningham & Kelly, 2017b).

We noticed that the teacher structured engineering activities and class discussions in ways that helped her pupils develop their facility to examine evidence and make decisions based in data. Students' participation in a common design challenge gave them a communal experience and a shared basis for deliberation. The teacher had students focus on a subset of relevant variables, and modeled for students how they should conduct tests. Because of this, students could share their data and use those data as a common basis for making engineering design decisions. Once the data were collected, the teacher guided discussions about patterned and anomalous data. She also structured the conversations so they focused on criteria related to engineering, which helped the students make connections between what they were doing and the work engineers do.

By making explicit her expectations for standard procedures, materials, and data collection techniques, the teacher further enabled class members to share and compare what they found and to learn from one another. For example, she asked her students to make predictions publicly and then invited the class to comment. Groups shared their data with the whole class—the data became common data. The teacher had the whole class analyze the shared data and draw conclusions. Because students shared their ideas and understandings publicly, all students were able to benefit from the shared insights to improve their next design. In this class, engineering knowledge was communal.

Finally, the teacher encouraged her pupils to develop agency as engineers. They collected data, designed solutions, shared out and analyzed their results, and then redesigned their technology. By sharing ideas and results publicly, they developed accountability to their peers and the class. Like professional engineers, the students were held accountable to the criteria and constraints set forth by the curriculum and the standards of their social group.

Developing Engineering Identity and Agency

As we observed classroom instruction, we were struck by the degree to which students identified and affiliated with engineering. Our interest in how teachers build student identity and agency with engineering led us to study in more depth how this can be fostered by classroom engineering instruction (Kelly et al., 2017). We watched videotape of classroom engineering instruction for two teachers and identified ways that interactions (between the students themselves and between the teachers and students) fostered an engineering identity over time.

One way that the teachers supported students' identities as engineers was by explicitly naming and referring to the students as "engineers" during various classroom moments. As students engaged in engineering work, their teachers called their work "engineering." In this way, the students' emerging "engineer" identities were connected to the work of engaging in engineering practice and talking about how such engagement entails being an engineer. The teachers made clear to their students that the class was involved in engineering.

Additionally, teachers used features of the engineering curriculum and results from experimental activity as tools for supporting students' understanding of and connection to engineering. They did this by making reference to the characters in the introductory storybook, connecting students' activities to the engineering design process, and using results of students' designs and experiments as a context for naming and taking up engineering practices.

Dr. Heidi Carlone's research group at University of North Carolina–Greensboro has studied the possibilities that engineering has for helping students reenvision their identities and abilities. They conducted research in low-income schools in North Carolina with a high percentage of Black and Hispanic students and explored how student and teacher ideas of "smartness" in the classroom may change as students and teachers begin to engage with EiE (Hegedus, Carlone, & Carter, 2014). Students who were considered "smart" in engineering exhibited different qualities (for example, persistence or creativity) from those considered "smart"—high achievers—in other subjects. This led some students who were generally low or moderate achievers in school to be considered "smart engineers," surprising their teachers and other students.

A research group led by Dr. Ann Robinson at the University of Arkansas at Little Rock explored the power of engineering (using EiE) to identify a more diverse group of students for participation in gifted-and-talented services (Robinson et al., 2018). The quasi-experimental study of 1st-graders from low-income schools compared students exposed to an engineering-enriched curriculum to a control group that received regular instruction. It examined whether engagement with engineering influenced the number of students, particularly those who were Black or Hispanic or who received a meal subsidy, who were nominated for gifted-and-talented services. It found that teachers who engaged their students in problem solving and engineering were more likely to nominate students, especially those from groups underrepresented in engineering. The authors concluded that a focus on creating problem solving and engineering design processes provides a promising alternate means for identifying students for gifted-and-talented services who might be otherwise overlooked.

Failure in Engineering

Failure plays a huge role in the engineering design process. Students and teachers may need support as they reconsider what failure means and how to learn from it. A few researchers are focusing their efforts on better understanding the role of failure in elementary engineering.

Dr. Matthew Johnson at Pennsylvania State University conducted a study examining videotape of students engineering bridges. He looked for where failure occurred during the activity and observed how students and teachers responded to these failures (Johnson, 2016). The research identified a variety of ways that student designs fail and how improvement might proceed. He categorized failure in classrooms along three continua: (1) intended or unintended (whether failure was intended or whether it occurred because the technology did not perform as expected), (2) high or low stakes (whether failure happened in small groups with the purpose of learning and revising versus in whole-class settings where students were showcasing their design), and (3) objective or subjective (whether the criteria were set externally or arose via comparison to others' designs). The research also diagnosed four causes for failure: (1) lack of knowledge of science and/or technology, (2) lack of understanding of materials and their properties, (3) poor craftsmanship, and (4) an inherent limitation in the materials available. The study suggested common obstacles that students might face as they try to improve their failed designs—lack of opportunity for students to improve their designs, lack of fair comparisons, and the use of unproductive improvement strategies such as basing decisions on misunderstandings or poor craftsmanship. The study also articulated three roles that teachers play as they support students through failed attempts. Depending on the context and needs at that time, teachers move back and forth between acting as a cheerleader, a manager, and a strategic partner. Helping students to expect and learn from failure is an important part of engineering. Understanding more about how students and designs experience failure can help educators shape instruction.

A second body of research conducted by Dr. Pamela Lottero-Perdue at Towson University and Elizabeth Parry, a STEM consultant, about failure in elementary engineering classrooms drew upon interview and survey data from students and teachers to investigate their responses to design failure. In a first study, Lottero-Perdue and Parry (2014) found that, although teachers may recognize that failure can be a learning opportunity, they do not use failure or "fail words" in the classrooms because of the traditional, negative connotations such terms often carry in education. In subsequent studies, these researchers categorized how students respond to design failure across two axes—identifying actions as resilient/productive or nonresilient/nonproductive and as positive or negative emotion and identities. They summarized the types of responses that teachers have to students with failed designs and created a model for these (Lottero-Perdue, 2015; Lottero-Perdue & Parry, 2015). Not surprisingly, their research finds that as teachers gain experience teaching engineering, particularly with curricula that scaffold students though failure, they become more comfortable supporting students through failed design attempts (Lottero-Perdue & Parry, 2016). The important role of failure in engineering means that we need to help teachers and students become comfortable with failure and with the types of responses they might experience. As engineering instruction affects both teachers and students, we turn now to examine briefly a few of the impacts of K–12 engineering on educators.

IMPACTS ON TEACHERS

Our model for supporting engineering education rests on three pillars—curriculum development, professional development, and research and evaluation. Teacher professional development (PD) is a critical element for the implementation of engineering in classrooms. There is a small but steadily growing number of preservice elementary teacher education programs that have begun to include engineering. These efforts are critical for preparing teachers. Our efforts have focused on inservice teacher professional development—there are currently more than a million elementary teachers in the U.S. and, for the vast majority of these, engineering is a new subject. We offer dozens of face-to-face and online workshops every year to introduce classroom teachers and teacher educators to engineering and to support their ongoing development as elementary teachers of engineering. Our approach involves introducing teachers to engineering by engaging them in engineering challenges as learners—we design the sessions so they model learner-centered, inquiry-rich learning. We also include reflective activities that ask PD participants to consider an engineering activity from a teacher's perspective to distill pedagogical strategies that facilitate classroom management and learning.

A question we continually ask ourselves is whether our PD offerings are meeting the project's goal of supporting teacher learning about engineering education and engineering implementation. To answer this question, our team collects a variety of data: feedback and other surveys from teachers, focus groups with teachers participating in our newest PD offerings, and interviews. We use these data to improve our sessions. For example, a number of participants suggested that instead of a second workshop activity focused on the engineering design process, they spend additional time on engineering facilitation strategies and pedagogies. We reorganized the workshop agenda to accommodate this preference.

In anonymous workshop evaluations, teachers consistently rate their PD experiences extremely highly, and tell us that they feel more prepared and confident in their understanding of engineering and how to implement engineering with elementary students. We have followed up with many teachers 6 months to a year after a workshop, to seek their perspectives after they have implemented engineering in their classrooms. Most teachers report that the PD prepared them well, and that they plan to teach engineering again the next year. They report positive impacts on their students' engagement, communication, and problem-solving skills. And, for themselves, many report that the experience of learning about and teaching engineering using EiE has influenced the way they teach. As one teacher informed us, "I have become more of a facilitator of learning rather than what you would call the traditional teacher. Students can put their own ideas together and learn from one another instead of my being the 'fountain of knowledge.'" Other teachers report the enjoyment they felt teaching engineering: "Using this program to teach my students reminded me of why I love to teach. It brought back the joy of teaching because my students were actively engaged in the learning process."

As part of our large-scale research study, we collected a Teacher Attitude survey from all teachers before they participated in an engineering professional development workshop and after 2 years of teaching engineering. We found that teachers became significantly more positive about engineering education in all areas after 2 years of participation in the study. They rated the relevance of engineering to their lives and students' lives more highly; they reported more use of project-based pedagogy in all their classes; and they reported higher enjoyment of science, math, and engineering. They also were more likely to agree that engineering should be taught at all grade levels, from elementary school to college. Most research study participants indicated that they planned to continue teaching engineering after the study was complete.

IMPACTS ON SCHOOLS

Schools that embrace engineering regularly allow all students to engineer. Such schools report gains not only in their students' science test scores but also in the students' problem-solving and design abilities. Engineering magnet schools in districts where school choice is available report that they become desirable choices for students and their parents. There are many ways that elementary schools and districts can incorporate or even structure their programs around engineering. Here, I profile two schools and describe some of the impacts they attribute to their engineering efforts.

Tully: Kids Who Engineer Outscore the District Science Average

In 2009, the University of Louisville J. B. Speed School of Engineering began an outreach program to K–12 schools in the region. Their goal: fostering kids' interest in engineering and eventually recruiting the students to study engineering with them. As the two leaders in this endeavor, Dr. Patricia Ralston, the chair of the Department of Engineering Fundamentals (which houses outreach programs), and Gary Rivoli, the director of outreach programs at the university, searched for available materials, they came across Engineering is Elementary. At the same time, EiE was building a national network of EiE "Hub Sites." Mr. Rivoli became part of the cohort, attended professional development to learn more about EiE, and received some funds to offer regional outreach. He approached Tully Elementary School's principal, who called a meeting with members of the school's faculty to discuss the K–12 engineering initiative. Fortunately for the school, Laura Keeling, then a 3rd-grade teacher, attended the meeting. She recalled: "They asked, 'Would anyone be interested in teaching this curriculum that's supposed to get kids interested in engineering?' We teachers were literally sitting on our hands, we were that intimidated!" Ms. Keeling felt especially hesitant:

> At the time, I was team-teaching with another 3rd-grade teacher and I used to *beg* her to teach the science so I could teach social studies! I was *not* scientific. I didn't have great science grades in high school, I wasn't interested in science during college . . . but since no one volunteered, my hand went slowly to the ceiling. I can't say why.

That summer, Ms. Keeling attended a professional development led by Mr. Rivoli. Once she taught her first engineering unit with her students, she was hooked. And a year later, when the STEM lab teacher retired, she "applied for the job. I thought, 'If I can do engineering with one class, surely I can do engineering with all 29 classes at the school!'"

In her new position, Ms. Keeling started small, teaching just one engineering unit to each grade, but kept building her program. Since the 2011–2012 school year, every one of the nearly 700 students in grades K–5 has engaged with the hands-on engineering activities in at least two EiE units each year, 4th- and 5th-graders have explored three units, and over the course of their academic careers, Tully students have been exposed to almost all 20 EiE units and the diverse fields of engineering they address, such as chemical, biomedical, and industrial engineering.

In the same year that Tully implemented engineering instruction schoolwide, the state legislature enacted a new assessment program, the Kentucky Performance Rating for Educational Progress, or K-PREP.

Tully's scores mirrored the findings of other researchers studying EiE—kids learned science better when they learned with EiE. They found this to be the case *especially* with the roughly 35% of the school population who are English learners, receive free and reduced-price lunch, or have educational disabilities (called "gap" students by the district). As Tully principal Linda Dauenhauer reported, "When the first scores came back, one of the biggest 'Ahas!' was that students identified as 'gap' did much better in science compared to other schools in the district." In 2011–2012, about 55% of students at Tully identified as "gap" scored "proficient" or "distinguished" in science, compared to just 45% in the district as a whole. That trend continued. In 2013–2014, 73% of these students scored "proficient" or "distinguished" in science, compared to 54% in the district as a whole and 68.5% statewide.

The benefits extended beyond test scores. Principal Dauenhauer regularly visited classrooms; during these visits, she heard about the impact of engineering on her students: "It's so nice when teachers tap me on the shoulder and point out a child who has been super quiet in the classroom before, or not really engaged, who's now highly engaged with the engineering activities."

She continued: "I think schools unfortunately have stripped so much creativity out of kids' hands . . . there is so much 'required' work to be done. But

EiE allows kids to exercise that creativity. When they walk into our STEM lab, you see smiles on their faces, and you know they are thinking, 'I can be a successful engineer!'"

Parents and principals have also recognized the value of engineering. Tully Elementary offers a number of tours for parents who are considering what school their children should attend. Tour guides highlight the school's engineering program because parents want their children to have access to it. Ms. Dauenhauer notes similar trends among principals: "Now other principals in our district and surrounding counties are reaching out to learn more about what we're doing with engineering instruction. I credit the way EiE engages students in higher-order learning." She also mentions that "EiE fits perfectly with the new Next Generation Science Standards [NGSS]," which Kentucky adopted in 2013. "So EiE has helped us to be ahead of the curve with NGSS." She continues: "I think that now that engineering is part of NGSS, a lot of people are scratching their heads, asking, 'How do we introduce this?' I think EiE is the answer to reaching the standards, and the testimony is really in the kids."

Brentwood: Engineering a School Turnaround

In Raleigh, North Carolina, not all residents are prospering from the knowledge economy that has developed in the region. In the school now called Brentwood Magnet Elementary School of Engineering, 75% of students come from economically disadvantaged backgrounds, while about 60% are Hispanic, and 40% speak English as a second language. At the school, 20 to 30% of students attained "proficiency" on state tests in 2008, marking Brentwood as one of the lowest-performing schools in Wake County.

In response, the district turned to a democratic solution: They surveyed parents and acted on what they heard. The bold new plan involved making Brentwood a magnet school focused on engineering. Not every teacher was happy or comfortable with the change, and those who preferred to transfer to another school in the district were given that option. The reorganized schedule was built around a 45-minute STEM block each day, and no students could be pulled out of class for things like special education services during this block. Thus, all students participated in all STEM activities. The STEM coordinator for the school, Emily Hardee, worked closely with Elizabeth Parry, the engineering educator who also conducted some failure research mentioned previously, to select curricula and plan professional development. Their chosen curricula included EiE for all students, with different units each year.

Ms. Hardee explained, "We liked that EiE was research based, designed to support learning for students of all backgrounds and abilities, and proven effective. . . . It also helped that the materials came in kits, since some teachers were uncomfortable with the idea of teaching engineering to elementary students." Another criterion the school used for choosing or creating STEM activities was their connection to real-world events.

Through the STEM coordinator (Hardee) and engineering educator (Parry), teachers received individualized assistance in the form of professional development and coaching. Periodic professional development workshops introduced pedagogical strategies, such as using STEM notebooks, and ongoing mentoring and co-teaching with Ms. Hardee helped teachers new to the school learn about engineering and build facility with the subject.

In moving engineering to the center of the curriculum, Brentwood translated the five-step engineering design process into a broader vision of the school's philosophy of learning. It named what "Brentwood engineers" do: "Ask critical questions, Imagine possibilities, Plan collaboratively, Create innovative solutions, Improve continuously." The engineering design process pervaded the school, and continues to do so. It can be found in bathroom stalls as well as the principal's office. Students called to the office for behavior issues use the EDP to scaffold a conversation in which they ask what the problem was, imagine other ways they could have handled it, and plan a course of action going forward. Using a structured problem-solving process invites students to take ownership of their behavior, just as they take ownership of their learning in academic settings.

Since reinventing itself, the Brentwood Magnet Elementary School of Engineering has seen steady gains in test scores. The growth in students' science scores has almost tripled—from 19 to 60% proficient—making it one of the top 10 elementary schools in the district. For other subjects, the school has met or exceeded the expected growth set by the state for each of the past 4 years. Such gains separate the school from its counterparts with similar scores. Four years ago, its performance was among the bottom four of the 110 elementary schools in the district—the improvements mean it has surpassed eight other schools in the rankings and expects to continue to move up.

The story of Brentwood's success looks even better when all the data are considered. Since the school adopted an engineering focus, learning by English learner students has greatly improved—because engineering activities are often hands-on and involve manipulating physical objects, students who struggle with verbal and written language can show what they know. In the most recent of the district's yearly STEM surveys, 5th-graders communicated a 13% greater understanding of engineering and its beneficial impacts on society than other students in the district. And 78% of Brentwood students expressed an interest in an engineering career, compared to 62% in other district STEM elementary schools. These numbers might be part of the reason that the Brentwood school is now a model for the district, state, and even nation—the school constantly entertains visitors who want to see and learn more about the engineering efforts. Brentwood's work has also brought the school accolades: Brentwood was named a Model STEM school (the highest level of recognition in North Carolina) in the fall of 2015—a designation awarded to only 12 schools in the entire state, with Brentwood representing the only elementary school chosen. The school has also received the Donald R. Waldrip Magnet School of Merit Award of Excellence from Magnet Schools of America, the second-highest honor given by the organization.

These examples highlight two ways that engineering can be introduced to bring about school change. In one case (Tully), change began at the teacher level, with the first steps of instructional leadership by a single teacher ultimately helping to transform a school. In the second case (Brentwood), school leaders were courageous enough to take parental input seriously and act on it. One brought in engineering through teacher leadership, one through administrative, and both achieved great results. The approach involved offering rich content to all students in an inclusive, hands-on fashion and tapping into the students' own voice and creativity. It placed trust in the teachers' capacity and professionalism in carrying out an ambitious change, and provided the needed professional development and support to make it successful. It included mechanisms to scale success. These are the ingredients of these dramatic success stories. Whether you're in a decision-making role for a school or district, or trying to make changes in your own classroom, you can use engineering to drive change.

CHAPTER 8

Conclusion

Engineering Engineering Education

Not long ago, I was displaying EiE materials at a conference for Massachusetts middle and high school students and educators. Three young women, perusing the booths, saw one of our display storybooks, and exclaimed, "I remember this! We did this. I loved this project." They continued to describe in great detail the two engineering units they had done 4 years ago. I recognized them immediately as three students that I had observed when their 4th-grade classroom engineered parachutes and membranes. I greeted each young woman by name—to their utter surprise—and asked how they were doing. Now in 8th grade, these three young women had recently won their school science fair, which resulted in the trip to the conference. All three were still interested in science and engineering and, clearly, doing well in the subject. Hearing them excitedly recall the details of their EiE engineering experiences drove home for me the lifelong impact that engaging engineering experiences can have on young students.

I've dedicated almost 2 decades of my life to achieving one personal and professional goal: to help *all* children become creative problem-solvers so that they can tackle the challenges of their future. My chosen solution: introducing engineering into elementary classrooms. Initially, I focused on developing exemplary resources. But as the project gained traction and success, I added another aim: to change education. I try to do this in 2 ways—by demonstrating new models for education and by guiding the work of others interested in creating inclusive, realistic curricular materials by sharing lessons learned, effective processes, and principles. As my team and I continue to pursue our goals, we approach the task like any engineer would—we use the engineering design process to engineer engineering education. Here's how.

ASK

My team and I started the curriculum design process by asking ourselves a series of questions. These questions helped us articulate *why* young students might benefit from engineering and *what* elementary engineering education might look like. We thought about what kids should know or be able to do. And we carefully considered ways to develop resources that attract, interest, and nurture the diversity of students who attend our schools today. We regularly revisit these questions, and as we extend our efforts, we pose many additional questions to further guide the *how* of our work:

- What resources, such as frameworks, structures, and guidelines, can we create to support teachers and districts as they adopt engineering?
- What are the advantages of engineering for students who are underserved and underrepresented in STEM?
- How can we modify existing resources and create additional tools to better support differentiated instruction? How can we better scaffold English learners as they engineer? How can we also do this for students who receive special education services?
- How can we use digital technologies to enhance engineering experiences and make them more inclusive?
- How might online professional development and online communities provide additional support to teachers and broaden the reach of PD? How can we best use online sessions to support materials-intensive engineering activities designed to foster work in teams and a classroom learning community?
- What are the impacts of engineering education?
- How do we conduct research to inform our work?

We cannot change the education landscape today without asking questions about how we can improve, how we can best serve all students. Reflecting on both the possibilities and the potential stumbling blocks we will face in the future shapes our decisions about where we should invest our energies as we work toward our goals. Similarly, teachers, schools, or districts should ask questions to define their purposes for engineering education and to identify opportunities and constraints that will shape their efforts. Such questioning also naturally prompts us to start generating ideas about new endeavors and initiatives.

IMAGINE

When my team and I started, we explored the many different ways we could teach engineering to young students. We brainstormed possibilities for years, generating thousands of ideas during many heated discussions. We debated the merits of having students generate their own engineering challenges versus having each unit specify one. We contemplated how much support to provide students as they worked through the challenge—how much should we guide their work? We wrestled with whether students should all choose from a specified set of materials or use anything they gathered from home. Ultimately, our decisions were guided by classroom teacher feedback. We worked closely with practicing teachers in racially, economically, and geographically diverse schools and districts. Our pilot teachers willingly tried new and risky activities, provided feedback, and grounded our lofty ideas. Only about 10% of our initial ideas made it into the final versions of our resources. We also learned how much testing we needed to conduct in our development laboratory to create curricular materials that would function in diverse but manageable ways; to determine criteria and constraints that allowed for open-ended solutions that were able to be achieved, but not immediately; and to develop testing procedures and rubrics for designs that resulted in measurable outcomes.

We needed to imagine what engineering might look like in the classroom—and then test our vision against the real world. Almost as soon as we had success with engineering in local classrooms, I proposed a larger goal of how to create an approach and resources that were scalable nationwide. We needed to set forth innovative thinking that was not so far removed from daily classroom life and teacher concerns that it did not resonate with practitioners. As we brainstormed, we kept this constraint at the forefront of our mind.

Fifteen years later, successes that we could only imagine at the beginning have come to fruition. We have designed models for developing inclusive, scalable curricula and shared these with the field. Engineering is now accepted as a field that can be taught at the elementary level. Through our work and that of others, engineering is included in the NGSS and many state standards. Children can be exposed to engineering in school and afterschool programs, during trips to local museums, during summer camps, on television programs, and through educational games.

We continue to brainstorm and innovate. I tell my team to expect that 50% of our ideas will not work well (or at all) the first time, but to remember that failure and iteration are necessary parts of the process. Team members need to be willing to take risks and think creatively within the constraints of the system to help push it forward in scalable and sustainable ways. For example, my team and I are exploring ways that video can support engineering education. We generate possible ways to create new digital resources—some that we can offer now, and some we might create in the future—to enhance students' learning or widen access. Some of our ideas are still too forward-thinking, and they remain a hopeful glimmer for the future.

As we work to bring engineering to all classrooms, we ponder new possibilities for differentiation and inclusion; new structures and assessments for classroom activities; and new supports, networks, and certifications that might be created for teachers. Curriculum developers and teachers also need to keep imagining possibilities for what engineering education can look like in the elementary setting.

PLAN

One EiE engineering unit requires approximately 8–10 hours of instruction. On average, the development of a unit involves a dozen staff, 30–75 educators, 4–10 cycles of revision, 3 years, and more than 3,000 hours of staff time! That's a lot of planning. You don't create a product of the size and impact of EiE without a scalability plan. The obstacles to and possibilities for the large-scale implementation we envision have to be sifted through, assessed, and verified before a model is proposed. As we develop our resources—curricular units, professional development, video supports, and so on—our planning process focuses on creating resources that can be used nationwide. We produce one structure and format for a curricular product that can be used in many different classrooms. To do so, we gather information, test our materials and products, and gather feedback from educators who work in a range of settings. Looking across the diverse stories and needs of our users challenges us to think about core elements and approaches that support student learning. In planning our resources, therefore, we must aim for a common design.

Additionally, we understand that each school and district is unique. We have worked with hundreds of schools and districts, and we have learned a lot about how to successfully implement engineering in school and district settings. When a school or district decides to implement engineering in its curriculum, we encourage teachers and administrators to create their own customized implementation plan and we often work closely with them. Understanding the local educational landscape—the goals, needs, available resources, existing systems, and limitations of the district or school—influences the plan that we create. We recommend that schools and districts discuss what they hope to accomplish, how they will know they are successful, and a plan for sustainability. We encourage leaders and teachers to consider how they will communicate with parents, other stakeholders, or the larger community about their new engineering projects.

CREATE

As of August 2017 our materials have been used by 172,000 educators to reach more than 14 million children nationwide; we designed them so they can be used by all. But we continue to explore ways that we can design additional supports to even better serve all students. We are currently developing additional modifications for English learners and students who receive special education services, translating our materials and professional development into Spanish, and creating digital storybooks. Demand from the field has also led us to develop engineering curricula for elementary and middle out-of-school settings and for kindergarten and pre-K students. Not only did we create a curriculum, but we were part of the effort to create national awareness of the need for engineering and the opportunities that making space for this new discipline in the school day can provide.

For schools and districts to create an environment supportive of a new field, they will likely need to address a few common hurdles. Funding is one of these. New curricula and teacher education both require investment. Fortunately, government funding agencies, industry, and philanthropic donors recognize the importance of problem solving, engineering, and STEM, and often are willing to underwrite some of the costs of introducing engineering with grants, gifts, or materials donations.

One tool that supports the introduction of a new school subject is curriculum. Another critical element is developing educators' knowledge and expertise through professional development (PD) for both inservice and preservice teachers. Soon after starting to create our classroom curriculum, we also begin to create PD sessions and resources. Carving out time to help teachers become familiar and comfortable with engineering concepts and practices as well as pedagogical strategies that can support student learning is important. Most districts, however, have limited time and resources for PD. So we work closely with districts and teachers to develop supports that help scaffold them through the introduction of engineering. We continue to develop a national network of professional development providers who offer face-to-face workshops and on-the-ground support. Evaluations of our PD regularly reinforce the importance of sessions that not only model instructional strategies but also give teachers an opportunity to engage in the activities they will teach. The feedback also highlights the value of providing time for reflection about the teacher-level moves that can shape high-quality engineering experiences. Recognizing the travel and time limitations of many teachers, we are also developing a range of online PD resources that support and connect teachers as they become more familiar with engineering and the principles of elementary engineering instruction.

Additionally, these online PD resources give teachers the opportunity to learn from their peers. Seeing engineering in action in a classroom and talking with peers who have already implemented a unit can be a powerful tool. We recommend that schools and districts begin their rollout with a visit to a school that has already taught engineering, to observe what students are capable of as they gather information and advice from the teachers and administrations about "lessons learned." Within a school or district, a set of early adopters might first tackle the integration of engineering to gain some familiarity and expertise. These leaders might then serve as a resource and support system as engineering is introduced more widely.

IMPROVE

Our development process is an iterative one. Elementary engineering education is still a relatively new field, and there is a lot we have yet to explore and learn. There is still so much work to do. Deep disparities persist in our schools and the educational opportunities afforded to students. Traditional models for STEM education are often uninteresting, irrelevant, inaccessible, or unavailable to many students. My team and I will continue to think about how to make education more widespread, equitable, and accessible using engineering as a tool. Often, we learn from teachers how to improve our processes and materials.

Recently, I was struck by the words of a talented 3rd-grade teacher from a high-poverty and extremely diverse school in Arizona. Occasionally, she invited her students to reflect upon their futures. "What do you want to *be* when you grow up?" she sometimes asked them. After teaching our engineering curriculum, she improved this question. She started asking instead, "What kind of *problems* do you want to solve when you grow up?" This is exactly the question we want students to ask themselves—an empowering question that leads to learning and commitment of the most constructive kind. It's the question that my team and I will continue to ask ourselves, and that I encourage you to consider as well.

References

Aikenhead, G. S., & Jegede, O. J. (1999). Cross-cultural science education: A cognitive explanation of a cultural phenomenon. *Journal of Research in Science Teaching, 36*(3), 269–287. Available at https://doi.org/10.1002/(SICI)1098-2736(199903)36:3<269::AID-TEA3>3.0.CO;2-T

Allie, S., Armien, M. N., Burgoyne, N., Case, J. M., Collier-Reed, B. I., Craig, T. S., . . . & Wolmarans, N. (2009). Learning as acquiring a discursive identity through participation in a community: Improving student learning in engineering education. *European Journal of Engineering Education, 34*(4), 359–367. Available at https://doi.org/10.1080/03043790902989457

Anderson, K. J. B., Courter, S. S., McGlamery, T., Nathans-Kelly, T. M., & Nicometo, C. G. (2010). Understanding engineering work and identity: A cross-case analysis of engineers within six firms. *Engineering Studies, 2*(3), 153–174. Available at https://doi.org/10.1080/19378629.2010.519772

Banilower, E. R., Smith, P. S., Weiss, I. R., Malzahn, K. A., Campbell, K. M., & Weis, A. M. (2013). *Report of the 2012 National Survey of Science and Mathematics Education*. Chapel Hill, NC: Horizon Research, Inc.

Blank, R. K. (2012). *What is the impact of decline in science instructional time in elementary school?* Noyce Foundation. Available at http://www.csss-science.org/downloads/NAEPElemScienceData.pdf

Blumenfeld, P. C., Soloway, E., Marx, R. W., Krajcik, J. S., Guzdial, M., & Palincsar, A. (1991). Motivating project-based learning: Sustaining the doing, supporting the learning. *Educational psychologist, 26*(3–4), 369–398.

Brotman, J. S., & Moore, F. M. (2008). Girls and science: A review of four themes in the science education literature. *Journal of Research in Science Teaching, 45*(9), 971–1002. Available at https://doi.org/10.1002/tea.20241

Brown, B. A., Reveles, J. M., & Kelly, G. J. (2005). Scientific literacy and discursive identity: A theoretical framework for understanding science learning. *Science Education, 89*, 779–802.

Bucciarelli, L. L. (1994). *Designing engineers*. Cambridge, MA: The MIT Press.

Buxton, C. A. (2010). Social problem solving through science: An approach to critical, place-based science teaching and learning. *Equity & Excellence in Education, 43*(1), 120–135. Available at https://doi.org/10.1080/10665680903408932

Carlone, H. B., Haun-Frank, J., & Webb, A. (2011). Assessing equity beyond knowledge- and skills-based outcomes: A comparative ethnography of two fourth-grade reform-based science classrooms. *Journal of Research in Science Teaching, 48*(5), 459–485. Available at https://doi.org/10.1002/tea.20413

Carlone, H. B., Lancaster, M. R., Mangrum, J., & Hegedus, T. (2016, April). *Fifth grade students' meanings of engineering competence.* Paper presented at the NARST Annual International Conference, Baltimore, MD.

Catsambis, S. (1995). Gender, race, ethnicity, and science education in the middle grades. *Journal of Research in Science Teaching, 32*(3), 243–257. Available at https://doi.org/10.1002/tea.3660320305

Cuevas, P., Lee, O., Hart, J., & Deaktor, R. (2005). Improving science inquiry with elementary students of diverse backgrounds. *Journal of Research in Science Teaching, 42*(3), 337–357. Available at https://doi.org/10.1002/tea.20053

Cunningham, C. M., & Carlsen, W. S. (2014a). Teaching engineering practices. *Journal of Science Teacher Education, 25*(2), 197–210. Available at https://doi.org/10.1007/s10972-014-9380-5

Cunningham, C. M., & Carlsen, W. S. (2014b). Precollege engineering education. In N. G. Lederman & S. K. Abell (Eds.), *Handbook of research on science education* (Vol. II, pp. 747–758). New York, NY: Routledge.

Cunningham, C. M., & Kelly, G. K. (2017a). Epistemic practices of engineering for education. *Science Education, 101*(3), 486–505. Available at https://doi.org/10.1002/sce.21271

Cunningham, C. M., & Kelly, G. K. (2017b). Framing engineering practices in elementary school classrooms. *International Journal of Engineering Education. 33*(1B), 295–307.

Cunningham, C. M., Lachapelle, C. P., & Davis, M. (in press). Engineering concepts, practices, and trajectories for early childhood education. In L. D. English & T. J. Moore (Eds.), *Early engineering learning.* New York, NY: Springer.

Duschl, R. A. (2008). Science education in three-part harmony: Balancing conceptual, epistemic, and social learning goals. *Review of Research in Education, 32*(1), 268–291. Available at https://doi.org/10.3102/0091732X07309371

Engineering is Elementary (EiE). (n.d.). The engineering design process. Available at eie.org/overview/engineering-design-process

Engineering is Elementary (EiE). (2007). *Hikaru's toy troubles.* Boston, MA: Museum of Science.

Engineering is Elementary (EiE). (2008). *Mariana becomes a butterfly.* Boston, MA: Museum of Science.

Engineering is Elementary (EiE). (2011a). *Sounds like fun: Seeing animal sounds.* Boston, MA: Museum of Science.

Engineering is Elementary (EiE). (2011b). *Now you're cooking: Designing solar ovens.* Boston, MA: Museum of Science.

Engineering is Elementary (EiE). (2011c). *The best of bugs: Designing hand pollinators.* Boston, MA: Museum of Science.

Engineering is Elementary (EiE). (2011d). *The attraction is obvious: Designing maglev systems.* Boston, MA: Museum of Science.

Engineering is Elementary (EiE). (2011e). *To get to the other side: Designing bridges.* Boston, MA: Museum of Science.

Engineering is Elementary (EiE). (2011f). *Thinking inside the box: Designing plant packages.* Boston, MA: Museum of Science.

Engineering is Elementary (EiE). (2011g). *A stick in the mud: Evaluating a landscape.* Boston, MA: Museum of Science.

Engineering is Elementary (EiE). (2016). *Shake things up: Engineering earthquake-resistant buildings*. Boston, MA: Museum of Science.

Faux, R. (2008). *Evaluation of the Museum of Science PCET project* (evaluation report). Somerville, MA: Davis Square Research Associates.

Fortus, D., Dershimer, R. C., Krajcik, J., Marx, R. W., & Mamlok-Naaman, R. (2004). Design-based science and student learning. *Journal of Research in Science Teaching, 41*(10), 1081–1110. Available at https://doi.org/10.1002/tea.20040

French, L. A., & Woodring, S. D. (2014). Science education in the early years. In B. Spodek & O. N. Saracho (Eds.), *Handbook of research on the education of young children* (pp. 179–196). New York, NY: Routledge.

González, N., Moll, L. C., & Amanti, C. (Eds.). (2006). *Funds of knowledge: Theorizing practices in households, communities, and classrooms*. New York, NY: Routledge.

Hegedus, T. A., Carlone, H. B., & Carter, A. D. (2014, June). *Shifts in the cultural production of "smartness" through engineering in elementary classrooms*. Paper presented at the American Society of Engineering Education Annual Conference & Exposition, Indianapolis, IN. Available at https://peer.asee.org/23013

Hill, A. M., & Anning, A. (2001). Primary teachers' and students' understanding of school situated design in Canada and England. *Research in Science Education, 31*(1), 117–135. Available at https://doi.org/10.1023/A:1012662329259

Hmelo-Silver, C. E., Duncan, R. G., & Chinn C. A. (2007). Scaffolding and achievement in problem-based and inquiry learning: A response to Kirschner, Sweller, and Clark. *Educational Psychologist, 42*(2), 99–107. Available at https://doi.org/10.1080/00461520701263368

Honey, M., Pearson, G., & Schweingruber, H. (Eds.). (2014). *STEM integration in K–12 education: Status, prospects, and an agenda for research.* Washington, DC: The National Academies Press.

Johnson, M. M. (2016, May). *Failure is an option: Reactions to failure in elementary engineering design projects* (doctoral dissertation). The Pennsylvania State University, State College, PA. Available at https://etda.libraries.psu.edu/catalog/28775

Jonassen, D., Strobel, J., & Lee, C. B. (2006). Everyday problem solving in engineering: Lessons for engineering educators. *Journal of Engineering Education, 95*(2), 139–151. Available at https://doi.org/10.1002/j.2168-9830.2006.tb00885.x

Kahle, J. B., Meece, J., & Scantlebury, K. (2000). Urban African-American middle school science students: Does standards-based teaching make a difference? *Journal of Research in Science Teaching, 37*(9), 1019–1041. Available at https://doi.org/10.1002/1098-2736(200011)37:9<1019::aid-tea9>3.0.co;2-j

Kelly, G. J. (2011). Scientific literacy, discourse, and epistemic practices. In C. Linder, L. Östman, D. A. Roberts, P.-O. Wickman, G. Erikson, & A. MacKinnon (Eds.), *Exploring the landscape of scientific literacy* (pp. 61–73). New York, NY: Routledge.

Kelly, G. J. (2014). Inquiry teaching and learning: Philosophical considerations. In M. Matthews (Ed.), *International handbook of research in history, philosophy and science teaching* (pp. 1363–1380). Dordrecht, The Netherlands: Springer. Available at https://doi.org/10.1007/978-94-007-7654-8_42

Kelly, G. J., Cunningham, C. M., & Ricketts, A. (2017). Engaging in identity work through engineering practices in elementary classrooms. *Linguistics & Education, 39*, 48–59. Available at https://doi.org/10.1016/j.linged.2017.05.003

Lachapelle, C. P., Cunningham, C. M., & Davis, M. (2017). Middle childhood education: Engineering concepts, practices, and trajectories. In M. J. de Vries (Ed.), *Handbook of technology education* (pp. 1–17). Cham, Switzerland: Springer International Publishing. Available at https://doi.org/10.1007/978-3-319-38889-2_23-1

Lachapelle, C. P., Hertel, J. D., Jocz, J., & Cunningham, C. M. (2013, April). *Measuring students' naïve conceptions about technology*. Paper presented at the NARST Annual International Conference, Rio Grande, Puerto Rico.

Lachapelle, C. P., Oh, Y., & Cunningham, C. M. (2017, April). *Effectiveness of an engineering curriculum intervention for elementary school: Moderating roles of student background characteristics*. Paper presented at the annual meeting of the American Educational Research Association, San Antonio, TX.

Lachapelle, C. P., Phadnis, P. S., Hertel, J. D., & Cunningham, C. M. (2012, April). *What is engineering? A survey of elementary students*. Paper presented at the 2nd P–12 Engineering and Design Education Research Summit, Washington, DC.

Lee, O. (2003). Equity for linguistically and culturally diverse students in science education: Recommendations for a research agenda. *Teachers College Record, 105*(3), 465–489. Available at http://www.tcrecord.org/Content.asp?ContentId=11114

Lottero-Perdue, P. S. (2015, April). *The engineering design process as a safe place to try again: Responses to failure by elementary teachers and students*. Paper presented at the NARST Annual International Conference, Chicago, IL.

Lottero-Perdue, P. S., & Parry, E. A. (2014, June). *Perspectives on failure in the classroom by elementary teachers new to teaching engineering*. Paper presented at the American Society of Engineering Education Annual Conference & Exposition, Indianapolis, IN. Available at https://peer.asee.org/22913

Lottero-Perdue, P. S., & Parry, E. A. (2015, June). *Elementary teachers' reported responses to student design failures*. Paper presented at the American Society of Engineering Education Annual Conference & Exposition, Seattle, WA. Available at https://peer.asee.org/23930

Lottero-Perdue, P. S., & Parry, E. A. (2016, June). *Elementary teachers' reflections on design failures and use of fail words after teaching engineering for two years*. Paper presented at the American Society of Engineering Education Annual Conference & Exposition, New Orleans, LA. Available at https://peer.asee.org/26923

Madhavan, G. (2015). *Applied minds: How engineers think*. New York, NY: W. W. Norton & Company.

Maltese, A. V., & Tai, R. H. (2010). Eyeballs in the fridge: Sources of early interest in science. *International Journal of Science Education, 32*(5), 669–685. Available at https://doi.org/10.1080/09500690902792385

Massachusetts Department of Education. (2001, May). *Massachusetts Science and Technology/Engineering Framework*. Available at http://www.doe.mass.edu/frameworks/scitech/2001/0501.pdf

Mathematics. (n.d.). In *Merriam-Webster's online dictionary*. Available at https://www.merriam-webster.com/dictionary/mathematics

Mayer, R. E. (2004). Should there be a three-strikes rule against pure discovery learning? *American Psychologist 59*(1), 14–19. Available at https://doi.org/10.1037/0003-066X.59.1.14

Mehalik, M. M., Doppelt, Y., & Schunn, C. D. (2008). Middle-school science through design-based learning versus scripted inquiry: Better overall science concept learning and equity gap reduction. *Journal of Engineering Education, 97*(1), 71–85. Available at https://doi.org/10.1002/j.2168-9830.2008.tb00955.x

Miaoulis, I. (2010). K-12 engineering—the missing core discipline. In D. Grasso & M. B. Burkins (Eds.), *Holistic engineering education* (pp. 37–51). New York, NY: Springer Science & Business Media. Available at https://DOI 10.1007/978-1-4419-1393-7_4,

Miaoulis, I. (2017, September–October). Engineering educational change. *Dimensions*, 40–43.

Miller, P. H., Blessing, J. S., & Schwartz, S. (2006). Gender differences in high-school students' views about science. *International Journal of Science Education 28*(4), 363–381. Available at https://doi.org/10.1080/09500690500277664

Minner, D. D., Levy, A. J., & Century, J. (2010). Inquiry-based science instruction—what is it and does it matter? Results from a research synthesis years 1984 to 2002. *Journal of Research in Science Teaching, 47*(4), 474–496. Available at https://doi.org/10.1002/tea.20347

Nasir, N. S., Rosebery, A. S., Warren, B., & Lee, C. D. (2006). Learning as a cultural process: Achieving equity through diversity. In R. K. Sawyer (Ed.), *The Cambridge handbook of the learning sciences* (pp. 489–504). New York, NY: Cambridge University Press.

NGSS Lead States. (2013). *Next Generation Science Standards: For states, by states*. Washington, DC: The National Academies Press.

Nightline (Director). (1999). The deep dive: Five days at IDEO. *Nightline with Ted Koppel*. New York, NY: ABC News.

Olitsky, S., Flohr, L. L., Gardner, J., & Billups, M. (2010). Coherence, contradiction, and the development of school science identities. *Journal of Research in Science Teaching, 47*(10), 1209–1228. Available at https://doi.org/10.1002/tea.20389

Partnership for 21st Century Skills (P21). (2009). *P21 framework definition*. Available at http://www.p21.org

Petroski, H. (2006). *Success through failure: The paradox of design*. Princeton, NJ: Princeton University Press.

Rhodes, M. (2016, April 27). Dyson's first-ever hair dryer will make all others look weak. *Wired*. Available at https://www.wired.com/2016/04/dyson-continues-take-home-making-hair-dryer/

Robinson, A., Adelson, J. L., Kidd, K. A., & Cunningham, C. M. (2018). A talent for tinkering: Developing talents in children from low-income households through engineering curriculum. *Gifted Child Quarterly*. Advance online publication. doi: 10.1177/0016986217738049

Roth, W.-M., & Lee, S. (2004). Science education as/for participation in the community. *Science Education, 88*(2), 263–291. Available at https://doi.org/10.1002/sce.10113

Roth, W.-M., & Lee, Y.-J. (2007). "Vygotsky's neglected legacy": Cultural-historical activity theory. *Review of Educational Research, 77*(2), 186–232. Available at https://doi.org/10.3102/0034654306298273

Sheppard, S., Colby, A., Macatangay, K., & Sullivan, W. (2006). What is engineering practice? *International Journal of Engineering Education, 22*(3), 429–438.

Silk, E. M., Schunn, C. D., & Strand Cary, M., (2009). The impact of an engineering design curriculum on science reasoning in an urban setting. *Journal of Science Education and Technology, 18*(3), 209–223. Available at https://doi.org/10.1007/s10956-009-9144-8

Sneider, C., & Purzer, S. (2014). The rising profile of STEM literacy through national standards and assessments. In S. Purzer, J. Strobel, & M. E. Cardella (Eds.), *Engineering in pre-college settings: Synthesizing research, policy, and practices* (pp. 3–20). West Lafayette, IN: Purdue University Press.

Trevelyan, J. (2010). Reconstructing engineering from practice. *Engineering Studies, 2*(3), 175–195. Available at https://doi.org/10.1080/19378629.2010.520135

Vasquez, J. A., Sneider, C., & Comer, M. (2013). *STEM lesson essentials, grades 3–8: Integrating science, technology, engineering, and mathematics.* Portsmouth, NH: Heinemann.

Vincenti, W. G. (1990). *What engineers know and how they know it: Analytical studies from aeronautical history.* Baltimore, MD: The Johns Hopkins University Press.

Weis, A. M., & Banilower, E. R. (2010). *Engineering is Elementary: Impact on historically-underrepresented students survey results.* Chapel Hill, NC: Horizon Research, Inc.

Wiggins, G., & McTighe, J. (1998). *Understanding by design.* Upper Saddle River, NJ: Merrill Prentice Hall.

Wulf, W. (1998). Diversity in engineering. *The Bridge, 28*(4). Available at https://www.nae.edu/Publications/Bridge/CompetitiveMaterialsandSolutions/DiversityinEngineering.aspx

Index

About the Author

Christine M. Cunningham is a vice president at the Museum of Science, Boston, where she works to make engineering and science more relevant, accessible, and understandable, especially for underserved and underrepresented populations. As the founding director of the groundbreaking Engineering is Elementary (EiE) project, she has developed engineering curricula for preschool through middle school students and professional development for their teachers. As of August 2017, EiE has reached 14 million children and 172,000 educators nationwide. Cunningham has previously served as a director of the Tufts University Center for Engineering Educational Outreach, where her work focused on integrating engineering with science, technology, and math in professional development for K–12 teachers. She also directed the Women's Experiences in College Engineering project, the first national, longitudinal, large-scale study of the factors that support young women pursuing engineering degrees. Cunningham is a fellow of the American Society for Engineering Education (ASEE). She has been awarded the ASEE K–12 and Pre-College Division Lifetime Achievement Award, the IEEE Pre-University Educator Award, the International Society for Design and Development in Education Prize, and the Alpheus Henry Snow Prize. In 2017 her work was recognized with the prestigious Harold W. McGraw Jr. Prize in Education. She holds BA and MA degrees in biology from Yale and a PhD in science education from Cornell University.